Getting Started with Adafruit Trinket

Mike Barela

D1531788

MAKER MEDIA™

SEBASTOPOL, CA

Getting Started with Adafruit Trinket

by Mike Barela

Published by Maker Media, Inc., 1005 Gravenstein Highway North, Sebastopol, CA 95472.

Maker Media books may be purchased for educational, business, or sales promotional use. Online editions are also available for most titles (*http://my.safaribooksonline.com (http://my.safaribooksonline.com/?portal=oreilly)*). For more information, contact our corporate/institutional sales department: 800-998-9938 or *corporate@oreilly.com*.

Editor: Brian Jepson and Emma Dvorak
Production Editor: Matthew Hacker
Copyeditor: Rachel Head
Proofreader: Carla Thornton
Interior Designer: David Futato
Cover Designer: Brian Jepson
Illustrator: Rebecca Demarest
Cover Photographer: Andrew Tingle

October 2014:   First Edition

Revision History for the First Edition

2014-09-25:   First Release

See *http://oreilly.com/catalog/errata.csp?isbn=9781457185946* for release details.

978-1-457-18594-6

[LSI]

Contents

Foreword. ix
Preface. xi

1/Introducing Trinket. 1
    Trinket Versus Arduino Uno. 2
    Using Trinket. 3
    The ATtiny85 Microcontroller. 3
    Memory. 4
    Connectivity. 5
    Three Volts or Five Volts?. 7
    The Adafruit Gemma. 8

2/Software Installation. 9
    Supported Operating Systems. 10
    The Preconfigured Arduino IDE from Adafruit. 10
    Modifying the Standard Arduino IDE. 11
    The USBtinyISP Driver for Windows. 11
    Seeing the Trinket in Windows. 14
    Windows Driver Troubleshooting. 15
    Linux. 15
    Conclusion. 15

3/Connection and Programming. 17
    Preparing the Trinket. 17
    Connecting Trinket to Your Computer. 19
    Loading a Program. 21
    The Trinket Data Pins. 24
        Digital Pins. 24
        Analog Pins. 25
    Exploring Data Pins. 25
        Parts List. 25
        Connections. 26
    Not All Pins Are the Same. 26

Different Ways to Power Trinket. 28
Analog and Digital Sensors. 30
Trinket Theremin. 30
    Parts List. 30
    Wiring. 31
    Code. 32
    Use. 34
Sound and Music. 34
Conclusion. 35

4/Libraries and Optimization. 37
Arduino Libraries. 37
ATtiny-Optimized Libraries. 39
Installing Libraries. 41
    Where to Install Libraries. 42
    Installing a Library in Windows. 44
    Installing a Library in OS X. 47
    Using Libraries. 49
Library Issues and Limitations. 50
Memory Optimizations. 50
    Program Space Optimization. 51
    Variable Optimization. 51
Conclusion. 53

5/Intermediate Projects. 55
Controlling Smart LEDs: NeoPixels. 55
    Important Things to Know About NeoPixels. 56
    NeoPixel Packaging. 57
NeoPixel Ornaments. 57
    Parts List. 58
    Build. 58
LED Color Organ. 61
    How It Works. 62
    Parts List. 62
    Build. 63
    Adjustments. 67
    Mounting. 67
Kaleidoscope Goggles. 68
    Parts List. 69
    Tools. 69
    Battery Selection. 70
    Wiring. 70

Software. 73

Final Assembly and Use. 75

Safety and Common Sense. 77

Servos. 78

Inside a Servo. 79

Trinket Servo Control. 80

Parts List. 81

Wiring. 81

Code. 82

Use. 83

Going Further. 83

Using I^{2}C—The Two-Wire Interface. 84

I^{2}C Software. 85

Using I^{2}C Displays. 85

Temperature and Humidity Sensing. 86

Parts List. 87

Libraries. 88

The LCD Display. 88

Testing the Display. 90

Adjustment. 91

Sensing. 91

Code. 93

How It Works. 94

Troubleshooting. 95

Going Further. 95

Ultrasonic Rangefinding. 96

Parts List. 97

Build. 97

Libraries. 99

Code. 99

How It Works. 102

Troubleshooting. 102

Communicating via Serial. 102

Talking Serial. 104

Exploring Serial Use. 104

Parts List. 104

Code. 105

Use. 106

Going Further. 108

Pulse Width Modulation. 108

The Analog Meter Clock. 110

Circuit Design. 111

Parts List. 111
Build. 112
Meters. 112
Libraries. 113
Code. 113
How It Works. 115
Preparing Your Meters. 116
Meter Mounting. 117
Conclusion. 119

6/Advanced Projects. 121
Trinket Jewelry. 121
Parts List. 122
Choices. 122
Tools. 123
Wiring. 123
Libraries. 124
Code. 125
Animation. 128
Compile. 131
Changing the Animation. 132
Finishing the Jewelry. 133
Program Memory for Data. 134
Trinket Occupancy Display. 135
Parts List. 136
Tools. 137
Wiring. 137
Libraries. 139
Code. 139
Enclosure and Board. 141
Box Connections. 142
Adjustment. 143
Room Placement. 145
Going Further. 145
Trinket Alarm System. 146
Parts List. 147
Tools. 147
Theory. 148
Multiple Sensors, One Pin. 149
Project Design. 151
Annunciation Selections. 152
Build. 153
Populating the Board. 154

Code.   157
Final Assembly.   160
Test.   161
Troubleshooting.   163
Going Further.   163
Bluetooth Communication.   164
Trinket Toy Animal.   165
Choosing Your Animal.   165
Parts List.   166
Tools.   167
Circuit.   167
Circuit Variations.   168
Code.   169
Preparing the Toy.   171
Use.   173
Trinket Rover Robot.   173
Parts List.   175
Tools.   175
3D Printing.   176
Build.   176
Wiring.   179
Code.   181
Going Further.   183
SPI Communications.   183
Trinket Audio Player.   185
Parts List.   186
Tools.   187
Software.   188
Loading Sounds.   188
Chip Loading Circuit.   189
Transferring Audio.   192
Sound Playback.   193
Use.   197
Conclusion.   197

7/Going Further with Trinket.   199
Microcontrollers: Smaller Versus Larger.   199
The Trinket Bootloader.   200
The Bootloader Design.   201
Bootloader Code.   201
Repairing the Trinket Bootloader.   202
Programming Bare ATtiny85 Chips.   203

Other AVR Programming Methods................................. 205

Community Resources... 206

    Learning Arduino.. 207

    Commercial Resources...................................... 207

    Technical Resources....................................... 208

    Third-Party Sites... 208

    Social Media Resources.................................... 209

8/Troubleshooting.. 211

Your USB Cable.. 211

Connectivity Issues.. 212

Arduino IDE Issues... 215

    Mac... 215

Common Library Problems...................................... 217

Error Messages... 219

    Compilation Issues.. 219

    Upload Errors... 221

    The Serial Monitor.. 222

Usage Issues... 222

Manufacturer Support... 224

A/Making Electronic Sounds...................................... 225

B/Parts Sourcing.. 231

C/Publications.. 235

Index... 237

Foreword

I like to talk about electronics and microcontrollers in terms of "BA" and "AA" (that's Before Arduino and After Arduino, by the way). In the days before Arduino, there were microcontrollers, to be sure. But it was really, really annoying to work with them. UV lamps, EEPROMs, one-time writes, high-voltage programmers! If you wanted to dabble in microcontrollers, the equipment and knowledge requirements were a steep hill to climb. Thanks to the beginner-friendly (but surprisingly powerful) Arduino, millions of engineers, artists, fashion designers, and more have been able to add electronics making to their skillsets.

At Adafruit, we've been doing Arduino projects for a very long time, and we've noticed that while some people like to push the capabilities to the very edge, there are many people who want something simple and small. A one-key keyboard, or an LED light-up brooch, or a servo driver. In many cases, the Arduino is great for prototyping but is a bit chunky. A smaller, simpler *mini Arduino* can do the job just fine. That's why we designed the Trinket, a miniature microcontroller board that can do little tasks nicely, and can be programmed similarly to the Arduino.

The Trinket builds on the great work of the Arduino team, including David Mellis, who first added Atmel ATtiny85 chip support to the Arduino development environment. Even though it may seem underpowered, there's something about the tiny size and simplicity of the Trinket (and its wearable sister, the Gemma) that inspires so many projects.

I'm delighted to introduce Mike Barela as the author of *Getting Started with Adafruit Trinket*. Mike has the deep engineering knowledge to explain the innards of a microcontroller or RC filter, the craftiness to detail how to build LED goggles, and the patience to line-by-line document the dozen projects in this book. As you read *Getting Started with Adafruit Trinket*, you'll find yourself immersed in the joy of hacking and figuring things out, learning how to tweak just a little bit more out of the little Trinket, while gaining knowledge of the same kinds of topics you'd run into with hulking 32-bit ARM processors.

Please try to build these projects, and—better yet—improve on them! Show them off to your siblings, parents, children, or friends. Give them as gifts, wear them to parties, and show off how much fun it is to Make!

—*Limor "Ladyada" Fried, Founder and Engineer, Adafruit*

Preface

The Trinket microcontroller provides designers with custom programmability in a size and price range perfect for modern projects. The number of projects using Trinket continues to grow, as witnessed in numerous project builds documented in social media. This book introduces you to some of the possibilities, providing a jumping-off point for your own explorations.

Who This Book Is For

This book is for *you*, the enthusiast who is expanding his knowledge of Making and controlling items through classes or self-study.

Working with Trinket is suitable for beginners, although it is assumed you have some familiarity with what a microcontroller is and with basic programming principles. The book steps through basic projects, working toward more challenging circuits and code. You'll find that I adapt and add concepts to create new functionality. After you complete the book, you can use it as a reference for microcontrollers, sensors, and coding techniques.

You will want to learn how to use the Arduino integrated development environment (IDE) for most of the examples in the book. Arduino compatibles are programmed in a variation of the C programming language with various prebuilt code in libraries. All of the code for the examples is supplied in the book and online (*http://bit.ly/GettingStartedWithTrinket*). For later projects, familiarity with electronics and project assembly is helpful but not required.

You will be following diagrams illustrating point-to-point wiring of electrical circuits. You'll be working on a solderless breadboard, which makes this easy to complete.

Recommended Reading

There is no required reading to work with this book, but here are some suggested resources that you may draw on to better understand particular subjects:

Component soldering with a soldering iron
> The book *Make: Learn to Solder* by Brian Jepson (Maker Media), Tyler Moskowite, and Gregory Hayes is a great reference. An alternative is the "Collin's Lab: Soldering (*https://learn.adafruit.com/collins-lab-soldering/*)" tutorial, which you can watch for free online. Both teach

the fundamentals of soldering, an essential skill for building electronic gadgets.

Familiarity with Arduino
The book *Getting Started with Arduino*, Second Edition, by Massimo Banzi (cocreator of Arduino), is a good resource, as is the Adafruit Learn Arduino series (*https://learn.adafruit.com/lesson-0-getting-started*), available for free online. Both offer an introduction to the Arduino open source electronics prototyping platform, including programming.

Other resources are listed in "Learning Arduino" on page 207.

What You Will Want to Have on Hand

To program a Trinket, you will need a Windows or Mac computer with a USB port. Linux may also work, although Adafruit does not guarantee compatibility with all Linux variants due to USB driver issues. Internet access is very helpful for obtaining example code, rather than typing it in yourself. The Internet is also great for reference material on specific subjects. The Adafruit Learning System (*http://learn.adafruit.com/*) and other websites post Trinket-related projects. Here are some other things you'll need to have on hand:

A good USB type A male-to-male Mini-B cable
I cannot stress this enough: get a *good* USB cable for programming the Trinket. Please consider buying a substantial USB type A male end to type Mini-B male cable, 3 feet (1 meter) long or less. So many visitors to Adafruit's Trinket support forum have repurposed old phone charging cables or other questionable cables. Such cables, more often than not, do not have the USB data wires required for communicating between the computer and the Trinket. Worn cables may work intermittently, but a good cable will save you hours of grief.

Basic tools

Soldering iron and solder
You will need a soldering iron and solder, to attach breadboard pins onto the Trinket and for building more permanent electrical circuits.

Multimeter
A multimeter capable of voltage and resistance measurements is a staple of any toolkit and can be purchased for under $10 or equivalent in most locations.

Pliers and wire cutters
Pliers and wire cutters are *essential*. Wires connect all the parts used in a project. Some other tools might be handy, too, including

a drill and screwdriver. Young makers may need assistance with a drill or other sharp tools that might be used in project packaging.

Electrical parts for projects

Most of the parts sourced in the book come from the manufacturer of the Trinket: Adafruit Industries (*http://www.adafruit.com*). To gather the parts you are looking for, turn to Appendix B. Adafruit has a good worldwide distribution system. Other hobbyist websites may provide similar parts, but you'll need more knowledge of how they work if they are not electrically the same as the specified parts.

 One advanced project uses 3D-printed parts. You can create these yourself (at home, at work, or in a Makerspace) or order them from a 3D printing service at a nominal cost.

Overall, working with the Trinket requires the same skills and materials as working with other hobbyist electronic items.

Conventions Used in This Book

The following typographical conventions are used in this book:

Italic

Indicates new terms, URLs, email addresses, filenames, and file extensions.

`Constant width`

Used for program listings, as well as within paragraphs to refer to program elements such as variable or function names, databases, data types, environment variables, statements, and keywords.

`Constant width bold`

Shows commands or other text that should be typed literally by the user.

`Constant width italic`

Shows text that should be replaced with user-supplied values or by values determined by context.

 This element signifies a tip, suggestion, or general note.

This element indicates a warning or caution.

Using Code Examples

This book is here to help you get your job done. In general, you may use the code in this book in your programs and documentation. You do not need to contact us for permission unless you're reproducing a significant portion of the code. For example, writing a program that uses several chunks of code from this book does not require permission. Selling or distributing a CD-ROM of examples from Make: books does require permission. Answering a question by citing this book and quoting example code does not require permission. Incorporating a significant amount of example code from this book into your product's documentation does require permission.

We appreciate, but do not require, attribution. An attribution usually includes the title, author, publisher, and ISBN. For example: "*Getting Started with Adafruit Trinket* by Mike Barela (Maker Media). Copyright 2015, 978-1-457-18594-6."

If you feel your use of code examples falls outside fair use or the permission given here, feel free to contact us at *bookpermissions@makermedia.com*.

Safari® Books Online

Safari Books Online is an on-demand digital library that delivers expert content in both book and video form from the world's leading authors in technology and business.

Technology professionals, software developers, web designers, and business and creative professionals use Safari Books Online as their primary resource for research, problem solving, learning, and certification training.

Safari Books Online offers a range of plans and pricing for enterprise, government, education, and individuals.

Members have access to thousands of books, training videos, and prepublication manuscripts in one fully searchable database from publishers like Maker Media, O'Reilly Media, Prentice Hall Professional, Addison-Wesley Professional, Microsoft Press, Sams, Que, Peachpit Press, Focal Press, Cisco Press, John Wiley & Sons, Syngress, Morgan Kaufmann, IBM Redbooks, Packt, Adobe Press, FT Press, Apress, Manning, New Riders, McGraw-Hill, Jones & Bartlett, Course Technology, and hundreds more. For more information about Safari Books Online, please visit us online.

How to Contact Us

Please address comments and questions concerning this book to the publisher:

Make:
1005 Gravenstein Highway North
Sebastopol, CA 95472
800-998-9938 (in the United States or Canada)
707-829-0515 (international or local)
707-829-0104 (fax)

Make: unites, inspires, informs, and entertains a growing community of resourceful people who undertake amazing projects in their backyards, basements, and garages. Make: celebrates your right to tweak, hack, and bend any technology to your will. The Make: audience continues to be a growing culture and community that believes in bettering ourselves, our environment, our educational system—our entire world. This is much more than an audience; it's a worldwide movement that Make: is leading—we call it the Maker Movement.

For more information about Make:, visit us online:

Make: magazine: *http://makezine.com/magazine/*
Maker Faire: *http://makerfaire.com*
Makezine.com: *http://makezine.com*
Maker Shed: *http://makershed.com*

We have a web page for this book, where we list errata, examples, and any additional information. You can access this page at:

http://shop.oreilly.com/product/0636920031598.do

To comment or ask technical questions about this book, send email to:

bookquestions@oreilly.com

Acknowledgments

I would like to thank Limor "Ladyada" Fried and Phillip Torrone of Adafruit Industries for the creative environment they have nurtured for Makers. Their encouragement and Adafruit's content donations are the foundation of the book. Adafruit's Frank Zhao and Phillip Burgess have built exquisite code and projects; their work, along with that of Adafruit contributors Rick Winscot and Bill Earl, has been incorporated with their permission and my thanks.

Thanks to the Maker Media team of Brian Jepson, Frank Teng, and Emma Dvorak, who guided the book's production.

Finally, thanks to Kate and Laura, who listened when presented with science fact and fiction. To Traci, your support and encouragement, across time and space, continues to make possible my many endeavors—my love, always.

1/Introducing Trinket

The Arduino has revolutionized the use of microcontrollers—programmable electronics —in the last several years, providing easy-to-use hardware and software at a reasonable price point. The often-cited Internet of Things has grown from this ubiquity of easy-to-use programmable electronics, sensors, and communications.

One of the few disappointments that typically comes after building a permanent project is, "I used my Uno in my project, and now I no longer have my $30 board." That, and the fact that many projects do not require all the horsepower and connectivity an Arduino Uno or larger board offers.

This "bigger is not always best" situation offered an opportunity to Adafruit Industries, a small company based in New York City. Specializing in innovative open source hardware, Adafruit has grown to become a premier supplier to hobbyists and industry. *Entrepreneur* magazine named Adafruit founder Limor "Ladyada" Fried as Entrepreneur of the Year for 2012, and she has been featured in *WIRED Magazine*, *Popular Mechanics*, and other publications.

Ladyada has an uncanny ability to look at the needs of customers and personally oversee the design of product solutions. The need for an inexpensive microcontroller that can be built into projects (without guilt) led to her introduction of the Trinket.

Trinket Versus Arduino Uno

As many people are familiar, at least in part, with the Arduino Uno, a comparison may help (see Figure 1-1).

Figure 1-1. *The Adafruit Trinket (left) and the Arduino Uno (right)*

Table 1-1 compares the features of each.

Table 1-1. *Trinket and Uno feature comparison*

	Adafruit Trinket	**Arduino Uno**
Pins (digital/analog)	5/3 (shared)	13/6
Pulse width modulated pins	3	5
Pin voltage	3.3 or 5 volts	5 volts
Memory (flash/RAM/ EEPROM)	8KB/512 bytes/512 bytes	32KB/2,048 bytes/ 1,024 bytes
Size (mm)	1.2 × 0.6 × 0.2 inch/31 × 15.5 × 5	2.96 × 2.1 × 0.59 inches/75.14 × 53.51 × 15.08
Approximate cost	$6.95	$29.95

Using Trinket

Many projects do not require the size, power, and capabilities of larger Arduino compatibles. Here are some categories of projects where the Trinket may be a good choice:

Wearables
> Wearables are a rapidly growing use for electronics. With its small size and low power requirements, the Trinket is being used in a growing number of clothing and body wear projects.

Sensing
> The Internet of Things is composed of many small smart sensors communicating information about the world around us. The Trinket is perfect for attaching a wide variety of sensors and displaying or communicating sensor status.

Tiny projects
> The Trinket is well suited for any use where programmability is desired in a small package. Very small robotics projects can be made with a Trinket.

Lights and display
> Coupled with light-emitting diodes (LEDs), the Trinket is a great choice for DIY lighting projects. Used with smart red-green-blue (RGB) LEDs, a Trinket can perform complex light animations. Adafruit's smart RGB LEDs, *NeoPixels*, are controlled with only one data pin. You can drive LED and character displays with only two pins.

New uses for the Trinket appear regularly on Internet project sites including Instructables (*http://instructables.com/*), Google+, and the Adafruit blog (*http://www.adafruit.com/blog/*), and forums (*http://forums.adafruit.com/*).

The ATtiny85 Microcontroller

At the heart of the Trinket is the ATtiny85 microcontroller (Figure 1-2), produced by Atmel Corporation. Despite having only eight pins in a tiny package, this controller provides the functionality of traditionally larger processors.

The ATtiny85 was introduced by Atmel as an extremely small controller on the outside with many of the features of larger processors inside.

Figure 1-2. *The ATtiny85 (the small black square on the Trinket)*

Memory

As you can see in Figure 1-3, this chip has three different types of memory. The ATtiny85 has 8,192 bytes of flash memory for programs. The Trinket contains *bootloader* code, which occupies part of this. The bootloader assists in loading user programs from the universal serial bus (USB) port. Adafruit has developed a custom bootloader based on the V-USB project (*http://www.obdev.at/products/vusb/index.html*). With the bootloader in flash memory, there is approximately 5,130 bytes of program memory available for user programs. Random access memory (RAM) is used for program variables. The ATtiny85 has 512 bytes of RAM, which seems like a minuscule amount compared to the 4 GB on a typical laptop, but in practice this is often more than enough for many programs.

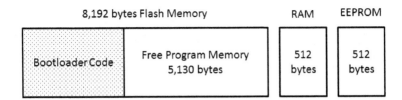

8,192 bytes Flash Memory		RAM	EEPROM
Bootloader Code	Free Program Memory 5,130 bytes	512 bytes	512 bytes

Figure 1-3. *The Trinket memory map*

Finally, the chip also contains 512 bytes of *electrically erasable program-mable read-only memory* (EEPROM). You can use this memory to store user data that remains even after the Trinket is powered off. This is useful to save data such as setup information, state data, or critical readings. This memory can also be useful for storing static information such as character strings a program might use, which otherwise would occupy precious program flash memory or RAM. However, programmers must weigh the benefits of using EEPROM against the additional code the compiler may add to manipulate data. Most programs do not use EEPROM.

Connectivity

The ATtiny85 chip uses only six pins for input and output, with two pins for power and ground. Atmel engineers cleverly assigned multiple types of functionality to each pin, as shown in Figure 1-4.

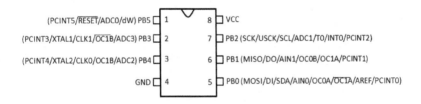

Figure 1-4. *The multiple functions on the ATtiny85 pins*

In the design of the Trinket, Adafruit exposes much of the chip functionality. The designers added the ability to communicate over the USB serial port, as well as status lights and a reset button. Figure 1-5 shows the Trinket 5V and the functionality onboard. The pins' functions are listed in Table 1-2.

Data is exchanged via the pins marked #0, #1, #2, #3, and #4. The sixth data pin (PB5) is permanently connected to the reset button and RST input; it cannot be used as an input/output pin due to how the Trinket is configured.

The Trinket has a power input pin, usually for a battery. There are also two voltage output pins: one for USB power (if connected to a computer) and a regulated power output tied to the battery input with a maximum power draw of 150 milliamps (mA).

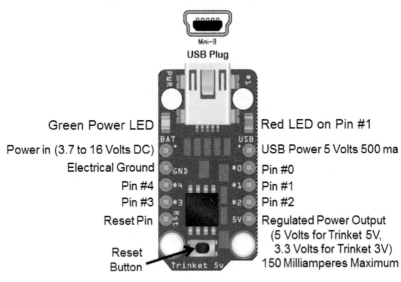

Figure 1-5. *Trinket connections*

Table 1-2. *Trinket pin use*

Trinket pin	ATtiny85 function	digitalRead digitalWrite	Analog analogRead	Pulse width modulation (PWM) analogWrite	I²C (TWI)	SPI	Pin is shared with
#0	PB0	0		0	SDA	MOSI	
#1	PB1	1		1		MISO	Red LED
#2	PB2	2	1		SCK	SCLK	
#3	PB3	3	3				USB D−
#4	PB4	4	2	With custom code			USB D+
RST	PB5	Connecting this pin to ground/GND resets the processor, similar to the reset button					Reset button
USB+		Provides 5 volts at 500 milliamperes when a USB plug is connected to the Trinket USB plug					USB power

Trinket pin	ATtiny85 function	digitalRead digitalWrite	Analog analogRead	Pulse width modulation (PWM) analogWrite	I²C (TWI)	SPI	Pin is shared with
BAT+	Battery power in (if not powered by USB), 3.7 to 16 volts (both Trinkets)						
GND	Electrical ground connection (negative power lead)						
3V or 5V	Provides 3.3 volts (Trinket 3V) or 5 volts (Trinket 5V) at 150 milliamperes						

Of particular note is that the ATtiny85 does not have native USB communication capability onboard. Adafruit wanted to develop a bootloader with the company's own USB identification. This allowed the design to use an existing computer driver Adafruit previously developed, which had the benefit of not requiring changes to the program that the Arduino software uses to transfer compiled code to the Trinket (avrdude).

Three Volts or Five Volts?

The Trinket comes in two versions. One operates at 5 volts direct current (DC), the other at 3.3 volts DC. The functionality of each is nearly identical. The 5-volt version can run from USB power or from an input voltage of 5 to 16 volts. The 3.3-volt version can run from USB power or an input voltage of 3.3 to 16 volts DC.

This provides a great deal of flexibility in powering a Trinket. A Trinket may be powered from a wall-mounted DC power supply (like a cell phone charger-type supply), of course. But it is also very suitable to being powered from a wide range of batteries. This includes batteries such as a single 3.7-volt lithium-polymer (LiPo) battery, three 1.5-volt batteries in series (4.5 volts), four batteries in series (6 volts for regular cells, 4.8 volts for rechargeables), or even a 9-volt battery (although a 9-volt may not provide current for a long time). The size of the batteries (the ampere-hour rating of the LiPo, or whether you use AAA, AA, C, or D cells) determines how long a circuit may last.

 Supply Voltage Designation

V_{CC} or VCC is one of the electronic designations for a project's voltage level. For this book, V_{CC} will generally be 5 volts for projects using a Trinket 5V and 3.3 volts for projects using a Trinket 3V or a Gemma (see "The Adafruit Gemma" on page 8).

The 3.3-volt version may be preferable when running off a 3.7-volt LiPo rechargeable battery. Sensors that operate on a 3.3-volt signal level are easier to use with a Trinket 3V. The only limitation the Trinket 3V has compared to the Trinket 5V is that the Trinket 3V can run at a clock speed of only 8 megahertz (MHz).

Many digital circuits operate at a signal level of 5 volts. Hooking a 5-volt circuit to a 3.3-volt input pin could damage the Trinket 3V's ATtiny85, so for projects that must use 5-volt signal levels, the Trinket 5V is the better choice. The Trinket 5V can run at a clock speed of 8 MHz or, via a software switch, at 16 MHz. Both the Trinket 5V and the Trinket 3V are used in projects in this book. You'll probably want to buy one of each for starters.

The Adafruit Gemma

The Adafruit Gemma (Figure 1-6) is a mini-microcontroller platform designed specifically for wearable projects. It contains the same ATtiny85 processor and bootloader as the Trinket 3V. Besides the easy-to-sew shape, the main difference is the addition of a premounted JST connector (white in the photo) to directly connect a LiPo battery. The Gemma does not expose data pins #3 and #4, as the Trinket does. More information on Gemma may be found at *http://learn.adafruit.com/introducing-gemma*.

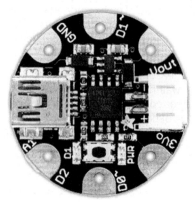

Figure 1-6. *The Adafruit Gemma*

If you need more than three data pins for a small project, the Trinket is a better choice than the Gemma. The Gemma comes only in a 3.3-volt version, whereas the Trinket has 3.3- and 5-volt options. Also, the Trinket 5V may be clocked to 16 MHz, twice as much as the Trinket 3V and Gemma. Code-wise, the Trinket and Gemma are identical.

2/Software Installation

The software used to program the Trinket is a modified version of the standard Arduino software. This is called the Arduino *integrated development environment* (IDE).

The Arduino IDE, shown in Figure 2-1 is the tool that has launched several million Arduino projects.

Figure 2-1. *The Arduino IDE window*

If you download the IDE from the official Arduino website (*http://arduino.cc*), it will not have Trinket programming capability installed. You can modify it to support the Trinket, or you can download a modified version of the IDE from Adafruit. Both options are covered in this chapter.

Supported Operating Systems

The Arduino IDE is supported on a number of operating systems (OSs). Adafruit officially supports use on Microsoft Windows (XP, Vista, 7, and 8) and Mac OS X. Microsoft's support for Windows XP ended in April 2014, limiting the security updates released from Microsoft. If you're considering Windows, use Windows 7 or higher (skip Vista if you can).

The Preconfigured Arduino IDE from Adafruit

Adafruit packages versions of the Arduino environment for Windows and Mac with all the necessary modifications for programming the Trinket, Gemma, and their larger wearable platform, Flora. The latest version of the software is available at *http://bit.ly/Trinket_Arduino_IDE*.

 Mac File Download

If you try to open the Mac version of the Arduino IDE (Adafruit's version or others) and the operating system says the file is damaged, corrupt, and needs to be trashed, it is most often due to stricter software security measures imposed by Apple in versions such as OS X Mavericks.

The preconfigured download package was not made by a "signed developer" and so is trapped by Mac OS X security. If you are using Mavericks or later, you will need to change the security setting to permit running of the Arduino IDE:

1. Go to System Preferences→Security & Privacy.
2. Click the lock icon and log in.
3. Change "Allow applications downloaded from" to "Anywhere."

You can now install the software. You should set the security setting back to the default once you've launched the preconfigured IDE for the first time. OS X will remember the setting for the IDE application even after you restore the security setting.

You can install multiple versions of the Arduino IDE on the same computer, but be aware that it will use several megabytes of disk space. Also, each version of the software will default to the same preference file, which may place all of a user's project files in the same location. You can change the location of the preference file by selecting the File→Preferences (Windows) or Arduino→Preferences (Mac) menu item; the location is the first entry in the dialog box. This will only affect the Arduino IDE in which you have made the change.

Modifying the Standard Arduino IDE

The modifications required to add Trinket support may change as Adafruit provides better integration with the Arduino IDE. If you are determined to apply individual changes to an existing Arduino IDE installation, see *https://learn.adafruit.com/introducing-trinket* for the most current step-by-step process. This includes adding a new *arduino.conf* and *ld.exe* linker, among other changes.

We recommend you download and install Adafruit's latest version of the Arduino IDE, which can still program "standard" Arduino boards but contains the enhancements needed for programming Trinket, Gemma, and Flora. The modifications you'll need to make to the standard Arduino IDE are numerous enough that missing one might result in more time spent debugging than simply installing the latest version.

Whether you've downloaded Adafruit's version or modified the standard Arduino IDE, you should now be ready to program the Trinket. If you are using Windows, however, you will need to install one more piece of software: the USBtinyISP driver.

The USBtinyISP Driver for Windows

For Mac and Linux, no driver is required. For Windows, you must download a USB communication driver called USBtinyISP. First, check which version of Windows you have by going to Control Panel and clicking the System icon, which will display your operating system version as shown in Figure 2-2.

Next, go to *http://learn.adafruit.com/usbtinyisp/drivers* and download the file for your version of Windows. Double-click the ZIP file you download and copy all the files you find into a new directory so you have an unzipped set of files for later.

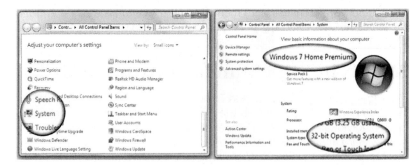

Figure 2-2. *Determining Windows version and type in Control Panel*

 Unzip That Driver!

Some folks have been frustrated to find that the USBtinyISP driver will not load. One reason this could happen is if you do not extract the files from the ZIP file you downloaded. Open the ZIP file with Windows Explorer and copy the files to another directory (such as a subdirectory under *Downloads*) before you install the driver.

Windows 8 and Higher

Windows 8 and higher ratchet up the security required for driver software. Starting with Windows 8.1, Microsoft implemented the requirement to install *signed drivers* on a routine basis. A signed driver is software to interface hardware with special *signature code* to authenticate the developer. This is intended to prevent you from installing malicious software. Arduino IDE versions 1.05 and 1.5 come with signed drivers for their boards. Adafruit has a signed driver for Trinket (which also works for Gemma and their USBtinyISP programmer). The direct link for this file is *http://www.adafruit.com/downloads/usbtiny_signed_8.zip*.

If you have followed the instructions at some point for loading an unsigned driver and you need to install the signed version, be sure to uninstall the unsigned driver before installing the signed driver.

If you are using Windows 8 or higher, be aware of this difference in drivers. The signed driver may also be used for Windows XP (32-bit), Windows Vista, and Windows 7.

Instructions for installing the driver manually are provided here for Windows 7 and 8. For Windows XP, see *http://learn.adafruit.com/usbtinyisp/drivers* for screenshots.

If you plug a new Trinket into a Windows computer without the USBtinyISP driver, you will see a driver installation message followed by a "device driver software was not successfully installed" message, as shown in Figure 2-3. If you do not get this message, press the reset button on the Trinket to have the device recognized as communicating with the PC.

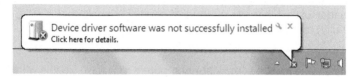

Figure 2-3. *Error message when Trinket is plugged into a Windows PC without the proper driver*

The manual method of installing the USBtinyISP device driver is as follows. (Reminder: If you have loaded a previous driver, it is best to unload that driver via Control Panel→Device Manager before installing a new one).

1. Plug your Trinket into your computer's USB port.

2. A USB 2 port (black plastic in the connector) is preferable. If you cannot use a USB 2 port, plug the computer's USB 3 port into a powered USB 2 hub, then plug the Trinket into the hub. If you have no other choice, you can try a USB 3 port (blue plastic inside the port connector). See Chapter 3 for connection options.

3. When your Trinket is connected to your Windows computer, go to Control Panel and search for Device Manager, then open it. You should see an unspecified device in the USB section.

4. Right-click and select Install Driver. Click the Browse button to select the directory with the unzipped USBtinyISP driver. Click Install. The driver should install, displaying the successfully installed dialog box (Figure 2-4).

Figure 2-4. *USBtinyISP driver installed correctly in Devices and Printers*

Seeing the Trinket in Windows

Although it is programmed via a USB cable, the Trinket does not act as a Windows or Mac serial communication device ("COM port"). This is because USB communication is not native to the ATtiny85. To work around this limitation, and to avoid the need for a separate USB chip, the Trinket bootloader uses a *bitbang* method of sending signals that a PC or Mac (hopefully) recognizes as USB.

 One reason some Linux computers may have trouble with the Trinket is that the Linux USB code may have timing issues with the USB protocol.

To confirm that Windows has correctly identified the Trinket:

1. Open Control Panel, and search for and open Device Manager.
2. Plug the Trinket into a USB 2 port via a cable to your computer.
3. Press the reset button on the Trinket. If Windows is set to make a sound on driver load, you should hear it now.
4. A new category should appear in the list: *libusb-win32 devices*. Under this you should see the entry *USBtiny*, as shown in Figure 2-5.

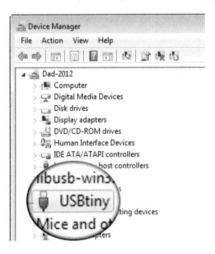

Figure 2-5. *Finding the driver in the Device Manager*

Right-click on USBtiny and select Properties to see if Windows believes the driver is loaded properly. The entry in Device Manager might disappear. This is fine. It should reappear if you press the reset button again.

Windows Driver Troubleshooting

If the Windows XP/7 driver does not want to load or has driver-related issues, try the signed Windows 8 driver. That driver version is compatible with Windows Vista, 7, and 8. It should also be compatible with Windows XP 32-bit, but not Windows XP 64-bit.

Other troubleshooting information is provided in Chapter 8.

Linux

As open source software (OSS) proponents, Adafruit originally supported the Trinket on Linux as well as Windows and Mac. User testing noted very strict USB port tolerances in the Linux input/output subsystem, however, which causes inconsistent communication results. You may program the Trinket on Linux, but Adafruit does not provide warranty support for those who do this.

The Adafruit tutorial on Trinket (*https://learn.adafruit.com/introducing-trinket/introduction*) is evolving as Linux issues are worked out. As of late 2014, Adafruit has developed specific versions of the Arduino IDE for Linux. Adafruit states that, as Linux compatibility is improved, the tutorial will be updated to reflect new information.

The utility that loads the program onto the Trinket, avrdude, typically uses superuser (root) privileges. You can configure Linux to not need these privileges, but the locations of the files you'll need to modify vary between distributions. See the latest tips for this in the Adafruit Trinket forum (*https://forums.adafruit.com/viewtopic.php?f=52&t=57062*).

Some machines (usually slower ones) may also require a change to *avrdude.conf*. See Chapter 8 for details.

Conclusion

Software evolves, and Adafruit works to improve the user experience with the Trinket and the software that works with it. This chapter got you set up with the IDE and drivers, which provide the interface to the Trinket. Next, the focus changes to hardware and code to allow projects to come together.

3/Connection and Programming

This chapter demonstrates connecting a Trinket to electronic components and introduces programming techniques. You'll build circuits and bring them to life with the code you write.

Preparing the Trinket

At this point, you'll need to gather the materials discussed in "What You Will Want to Have on Hand" on page 12.

Most experimenters use an electronics *breadboard* to mount their projects. These solderless boards (Figure 3-1) come in many sizes. The half size or even the mini size works well for Trinket projects. A breadboard allows the experimenter to mount a Trinket and have easy access to make connections to other components.

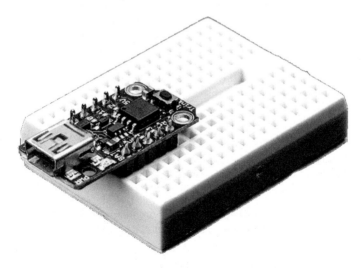

Figure 3-1. *Trinket with soldered headers on a breadboard*

You'll need a soldering iron to connect *headers* to the Trinket board. Headers, as shown in Figure 3-2, give you a secure connection between circuit boards and external components such as cables, other circuit boards, solderless breadboards, and wiring. Having plenty of headers in your parts collection is a good practice.

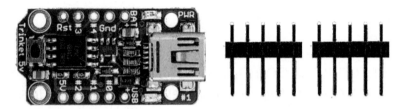

Figure 3-2. *Header pins make working with breadboards easier*

Trinkets come from Adafruit with some male headers for breadboard use. Your header may not come in two precut pieces of five; if this is the case, you will need to cut the header to create a row of five for each side. Firmly secure the header while cutting on the divide between pins. Hold both ends of the header so one end doesn't go flying. Eye protection is strongly recommended. Once you have the two rows of five, you can solder them most easily by pushing the rows into a breadboard and placing a Trinket on top. The long pins should be in the breadboard, with the shorter pins facing up toward the Trinket.

Working with a hot, clean soldering iron and a roll of solder, carefully apply the tip of the iron onto the area where the header pin touches the gold of the Trinket pin pad. Wait a second, then place the solder onto the joint area; it should flow around the metal. Remove the iron. There should be a nice, shiny, silver-looking coating on the joint. If it blobs, looks grey, or does not otherwise flow, make sure the iron is hot and the soldering tip is clean (before soldering, apply a tiny bit of solder to the tip to *tin* it). Repeat the process, soldering all 10 pins (Figure 3-1 shows the Trinket with all the pins soldered). When cool, you can remove your Trinket—it is now ready for use in experimenting.

For the rest of the book, you will need various electrical components. For example, for making circuit connections, you'll need either solid hookup wire that you cut and strip, or premade breadboard wire. Most electrical items sourced in the book come from Adafruit Industries (*http://www.adafruit.com*), but many are also available at Maker Shed (*http://www.makershed.com/collections/parts-components*) or your favorite electronics supplier. See Appendix B for a list of parts suppliers.

One type of component you will need that is not called out by Adafruit part number in this book is *resistors* (Figure 3-3). A fundamental electrical component, resistors impede current flow and thus are indispensable in a majority of electrical projects. You can buy them in individual resistance values, usually in packs of five or more, at your local electronics outlet. In the United States, RadioShack is a good source for these, and see Appendix B for some other parts suppliers. Some suppliers sell a package containing several of each of the most common resistor values. This may be a more practical and economical buy. These packages are also available at many parts outlets, including Maker Shed, RadioShack, and other sources.

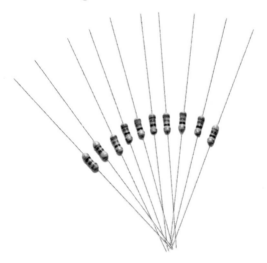

Figure 3-3. *Assorted resistors*

If you are able, you can also use advanced electronic testing equipment. It's not a requirement, but it is often handy to view signals with an oscilloscope or logic sensor. If you have access to such devices, they could help you find some obscure issues with circuits. Some schools and Maker spaces have this equipment available.

Connecting Trinket to Your Computer

A Trinket works best connected to a USB 2 port (the ones with the black colored plastic piece in the connector on your personal computer). If you do not have any USB 2 ports, you can use a USB 3 port (with blue colored plastic in the connector), but the electrical signal timings on some USB 3 ports may not work well with a Trinket.

One way to gain a USB 2 port is plugging a USB 2 hub into the computer's USB 3 port (Figure 3-4, bottom). This will typically provide several USB ports. To ensure you have plenty of power, I suggest a powered hub such as the one stocked by Adafruit (*http://www.adafruit.com/products/961*). If your computer works with the built-in USB port, the hub is not required: it is just noted as one solution for those who encounter issues.

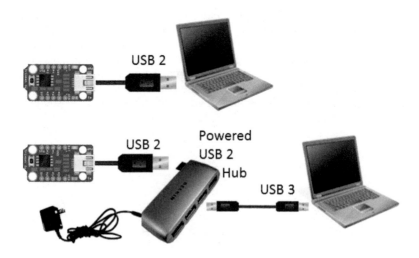

Figure 3-4. *Connecting Trinket via USB 2 directly or via a USB hub*

Plug the male USB A side of the cable into the computer and the male USB Mini-B side into the Trinket. If connecting via a hub, connect the computer to the hub, plug in the hub if necessary, and then connect the hub to the Trinket.

You should notice the green LED on the Trinket light up and the red LED blink brightly for the first 10 seconds it is plugged in. If you have no lights, there is a misconnection or bad cable.

 If no LEDs light up when you apply power, remove power and check your connections. If both LEDs light up but the red LED only turns on dimly for a second or two and goes out, the USB data lines were not detected properly. Try a new USB connection and be sure there are no extra connections on pins #3 and #4.

If your red LED continues to blink, that is perfectly fine; Adafruit tests its boards at the factory, and the blinking is from the test program.

 Never, ever connect a Trinket to wall (mains) power directly on any pin. Blue smoke and fire will probably result. If working with mains power, use an appropriate DC power supply and peripherals such as a PowerSwitch Tail to do switching.

Once everything is connected correctly, you can download your own programs to your Trinket.

Loading a Program

Arduino programs are called sketches. One of the best programs to test a Trinket is the ubiquitous Blink sketch. Our Blink sketch will be a bit different, both to test the Trinket's capabilities and to be sure it is our program running and not the factory test sketch. Example 3-1 shows our Blink sketch with each line of code numbered and explained.

Example 3-1. Our first program to test a Trinket

```
/*
 Blink
   Turns an LED on for one half second, then off for one half second,
   repeatedly.

   To upload to your Trinket:
   1) Select the proper board from the Tools->Board menu.
   2) Select USBtinyISP from Tools->Programmer.
   3) Plug in the Trinket, make sure you see the green LED lit.
   4) Press the button on the Trinket - verify you see
      the red LED pulse. This means it is ready to receive data.
   5) Click the IDE upload button above within 10 seconds.
*/
int led = 1;         ❶

void setup() {       ❷
  pinMode(led, OUTPUT);   ❸
}

void loop() {        ❹
  digitalWrite(led, HIGH);   ❺
  delay(500);        ❻
  digitalWrite(led, LOW);    ❼
  delay(500);        ❽
}
```

❶ Define which digital pin to write to. For the Trinket this would be pin #1, which is connected to the onboard red LED.

❷ The setup routine runs once when you power on or press the reset button.

❸ This command sets the LED pin (#1) as an output.

❹ The loop routine runs over and over again.

❺ Setting the LED pin as HIGH turns on the current to the LED, lighting it up.

❻ The delay function delays the number of milliseconds (thousandths of a second). In each delay we wait 500 milliseconds, which is half a second.

❼ Setting the LED pin as LOW turns it off.

❽ Blinking the LED twice a second helps distinguish this program from the once-a-second flash when the bootloader runs.

To ensure the program will properly *compile* (be converted from source code into machine code) and load onto the Trinket, perform the following steps in the Arduino IDE:

1. Select the Adafruit Trinket 8 MHz board from the Tools→Board submenu, as shown in Figure 3-5.

2. Next, select USBtinyISP from the Tools→Programmer submenu (see Figure 3-6).

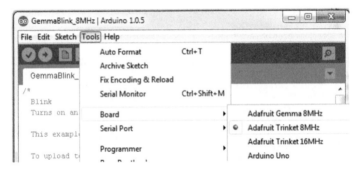

Figure 3-5. *Select the Trinket as your desired board in the Arduino IDE*

 Check It Twice

You need to select the correct board type and programmer, as in Figure 3-5 and Figure 3-6. This is crucial, but easy to overlook. Check these settings frequently if you can't otherwise resolve error messages. This could save time and tears.

Plug the Trinket into the USB cable. Make sure you see the green LED lit (power good) and the red LED pulsing. Press the tiny Trinket reset button if the red LED is not pulsing. When the red LED blinks, the Trinket is in *bootloader mode* (waiting to accept the program download).

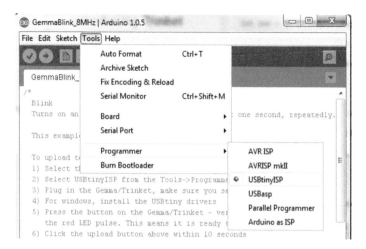

Figure 3-6. *Select USBtinyISP as the programmer for the Trinket*

Click the upload button (or select File→Upload), as shown in Figure 3-7.

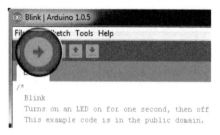

Figure 3-7. *The Arduino IDE upload button*

If everything goes smoothly, you should see the screen shown in Figure 3-8 (note no red error messages). The red LED on the Trinket will blink on and off once a second, as we have now programmed it to do.

The bottom part of the window will display how many bytes of program flash memory your program used, out of all the memory available (around 5,310 bytes).

If you get error messages, check for programming errors. If you have non-programming errors, see Chapter 8 for troubleshooting tips.

Figure 3-8. *Success! The program is compiled and loaded (no red errors in the black area)*

The Trinket Data Pins

There are five data pins on the Trinket, numbered on the board #0, #1, #2, #3, and #4. They each have multiple uses, depending on how you program it. Not all functions can be programmed on all pins, due to how Atmel designed the ATtiny85 chip and how Adafruit has designed the Trinket.

Digital Pins

A digital signal is either on or off, electrically high or low. What voltage high means depends on the Trinket model: for Trinket 5V, high is about 5 volts and low about 0 volts, and with Trinket 3V, high is about 3.3 volts and low about 0 volts. If you dig deeper, there is a bit of wiggle room in these values —typically a high is detected at more than one-half the supply voltage and low at less than half. But to give a clean digital signal, circuits are designed to get as close to zero and V_{CC} as possible to avoid false readings.

All five data pins may be digital inputs or outputs. For inputs, we can read from these pins with a high or low reading, and for outputs, we can output a high (supply voltage) or low (no voltage).

The Blink program uses a digital output. Digital pin #1 is designated as an output in setup with the pinMode function. The pin is then set first high then low once a second via the digitalWrite function calls. Trinket pin #1 is the only pin that we can control and see results from without needing an external component. That's because Trinket has an onboard component wired directly to it: a red LED and a current-limiting resistor that sets the amperage flowing from the ATtiny85 through the LED to ground at about 10 milliamps (0.010 amperes).

Analog Pins

As for the analog pins, they are very handy for reading a voltage and acting on the reading. See how to use them in "Trinket Theremin" on page 30.

 When using an analog pin (A1, A2, or A3) with a regular Arduino, you would typically refer to it in your code with the constants A1, A2, or A3. But with the Trinket, you need to use the numbers 1, 2, or 3 instead of A1, A2, or A3 due to an Arduino IDE bug. So, a call to analogRead(A1); on an Arduino Uno would be written analogRead(1); in a sketch for a Trinket. The compiler will not confuse this with D1, such as in digitalWrite(1);, because you are using an analog function.

Exploring Data Pins

An LED can be connected to any of the Trinket data lines. The circuit can blink an LED, similar to blinking the built-in LED as we did earlier. You can set this up on a breadboard with a resistor and an LED as shown in Figure 3-9 (right). You'll then be able to blink the LED from whatever data pin you connect it to.

Parts List

- 470-ohm resistor. The colors will be yellow-purple-brown, reading from the closest end to the wire. The last band will be silver (10%), gold (5%), or no additional color (20% stripe).
- An LED, any color you have on hand. The diode has a polarity, as shown in Figure 3-9 (left). The side with a flat part and a short lead is

the negative (cathode) lead, and the other side with the long lead is the positive (anode) lead.

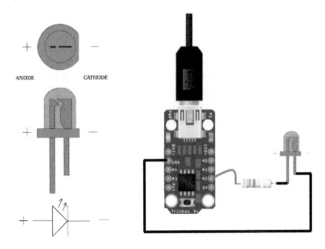

Figure 3-9. *LED polarity and connecting a resistor to an LED and to the Trinket*

Connections

The resistor is connected from pin #2 to the LED anode and the LED cathode is connected to ground. Ground should be the GND pin on the Trinket. Change the Blink program to declare the new LED pin to be pin #2 by changing `int led = 1;` to `int led = 2;`. Upload the new program, and the LED should blink just like the red onboard Trinket LED did when you ran the earlier Blink example. If there are problems, first check your program, then the wiring. Check your LED; you may have accidentally reversed it. The LED will not burn out on the Trinket if you reverse the leads, but you should avoid connecting LEDs backward in other circuits in general.

You can do this for other pins, but if you choose pin #3 or #4, be sure the Trinket is disconnected from the circuit while programming. Why are these pins different? This is explained next.

Not All Pins Are the Same

Due to the Adafruit implementation of connecting the Trinket to the USB for programming, and the red LED on pin #1, there are extra components on some of the ATtiny85 pins to watch out for. Figure 3-10 shows how the pins are connected internally.

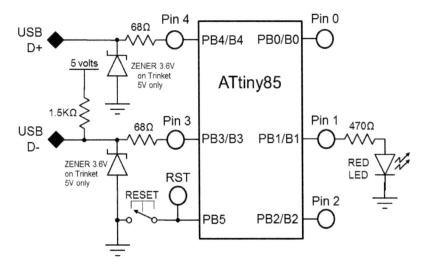

Figure 3-10. *ATtiny85 to Trinket pin connections*

Pins #3 and #4 are used when the bootloader is active on the USB connection. These pins have resistors between them and the USB connector to make their signals compatible with USB specifications. Pin #1 has the resistor and red LED combination onboard. The reset (RST) pin is not usable as a general-purpose pin. You can activate the reset pin by pressing the tiny reset button on the Trinket, or by connecting the RST pin to ground.

Despite the added components, pins #1, #3, and #4 can still be used for input and output on the Trinket, the same as pins #0 and #2. Electrically, components or methods may have to change when connecting to pins #1, #3, and #4 to get the results desired. The projects in the book will tell you when these changes are necessary.

There are some differences in how pin #1 and, curiously, pin #4 react electrically to what are called internal *pull-up resistors*. Pull-up resistors are similar to the 1,500-ohm resistor on pin #3 in Figure 3-10. The use is to "pull" the voltage on a pin toward the supply voltage (V_{CC}, 5 volts for Trinket 5V, 3.3 volts for Trinket 3V). External circuitry can "pull" the voltage down toward zero or ground, like the reset button does when it's pushed. The ATtiny85 has *internal* pull-ups, valued at a nominal 22,000 ohms, which you can set with pinMode(*pin*, INPUT_PULLUP);. Many developers count on this feature in their designs. Pin #1 has a 470-ohm external resistor and an LED on the Trinket, pulling that pin toward low. The internal pull-up does not provide enough balance to pull the pin high. Now for the strange, undocumented behavior: pin #4 also will display neither a digital high (V_{CC} signal level) nor low (a zero signal level). The value that results is about 2.16 volts when the internal pull-up is activated. This ambiguity (not a digital high or low reading) is not desirable in digital circuit design. The solution is to place an external 1,000-ohm resistor between the pin and the supply voltage (USB+, BAT+, or the 5V or 3V pins). The resistor will provide enough of a current path to pull up the pin in the manner digital circuits expect.

For complete schematics (electrical drawings) of the Trinket 3V and 5V, see *http://learn.adafruit.com/introducing-trinket/downloads*.

Different Ways to Power Trinket

So far, our Trinkets have been powered via the USB connection from a computer, as in Figure 3-11. This works for small circuits, up to a combined current draw of 500 milliamps (0.5 amps) on a standard USB 2 port. But the cable tethers the circuit to the computer, something not desired for most projects for a long period of time.

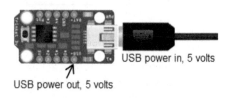

USB power in, 5 volts

USB power out, 5 volts

Figure 3-11. *Powering Trinket via a USB cable*

Fortunately, the Trinket is very flexible when it comes to getting power. The BAT+ connection shown in Figure 3-12 can take 3 to 16 volts for Trinket 3V and 5 to 16 volts for Trinket 5V.

Wall / mains power
5 to 16 volts DC
step-down transformer

Figure 3-12. *Using a DC power brick with Trinket*

You can also use the BAT+ terminal to connect a battery. There is a wide range of batteries from which to choose. A lithium polymer (LiPo) battery like the one shown in Figure 3-13 is just one choice. Battery packs made of AAA or AA batteries also work well. AA or larger batteries can provide current for longer periods than smaller batteries. Battery packs may connect to the BAT+ terminal or, on version 1.1 Trinkets, to the back with an optional surface-mount JST connector (Figure 3-13, right).

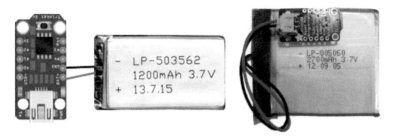

Figure 3-13. *Connecting a LiPo battery to a Trinket 3V*

 Rechargeable AA and AAA batteries supply slightly less power (1.2V) than their alkaline counterparts (a little over 1.5V when fresh), so factor this in when calculating how many batteries to use.

The 5-Volt Trinket's Little Secret

The Trinket 5V is actually comfortable running below 5 volts, down to 3.3 volts. The Trinket IDE board type (in Tools→Board) should only be set for Adafruit Trinket 8 MHz at supply voltages lower than 4.5 volts. Above 4.5 volts, you can set the board type to Adafruit Trinket 8 MHz or Adafruit Trinket 16 MHz in the IDE. The ability to power a Trinket 5V at a lower voltage allows it to be connected to a LiPo battery at 3.7 volts, for example. Note, however, that if the voltage to BAT+ is less than 5 volts, the voltage out of the 5V pin will not be 5 volts, and the signal pins will be operating at 3.3 volts.

As an added power option, Adafruit released a slight revision to the Trinket in 2014. The back was redesigned to allow soldering an optional surface-mount JST connector (Adafruit part #1769), shown in Figure 3-14. JST is the connector type used on many batteries and battery packs. This is a very welcome addition, and the connection brings Gemma-like power simplicity to the Trinket.

The versatility of powering Trinkets provides flexibility in designing small, powerful projects.

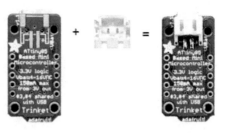

Figure 3-14. *Using a JST connector on newer versions of Trinket*

Analog and Digital Sensors

Once you can blink LEDs, you'll want to start connecting something interesting. Most projects contain some type of sensor to provide the ability to sense the world around us. The Trinket is ideal for connecting a sensor, processing its input, and reacting to or communicating the results.

Trinket Theremin

The next project is rather fun. Some background: Léon Theremin was a Russian inventor who created one of the first electronic musical instruments. His electronic instrument, also called the theremin, used the idea of electrical circuit resonance. The circuit here differs a bit but still keeps the magic of waving your hands to make music.

This project combines a light-sensitive cadmium sulfide (CdS) photocell with a Trinket and a piezo speaker to play music.

Parts List

- Trinket 5V, Adafruit #1501
- Half breadboard, Adafruit #64
- Breadboard jumper wires, Adafruit #153
- Cadmium sulfide (CdS) photoresistor, Adafruit #161
- Piezo buzzer/speaker, Adafruit #160
- Female 5.5/2.1 mm DC power adapter, Adafruit #368
- 5V, 2A power supply, Adafruit #276 or similar
- 1,000-ohm resistor

No tools are needed beyond those required to solder headers onto a Trinket for breadboard use, as noted in Chapter 2.

Wiring

The connections diagram is shown in Figure 3-15. This type of diagram is made with the program *Fritzing*, available for free at Fritzing.org (*http://www.fritzing.org/*). It has become popular for drawing project connections. Adafruit makes Fritzing parts for many of its designs, including the Trinket, available in a downloadable parts library at *https://github.com/adafruit/Fritzing-Library*.

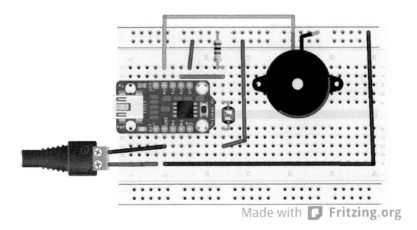

Figure 3-15. *The connection diagram for the Trinket Theremin*

The Trinket pin connections are as follows: pin #0 (digital pin 0) is connected to the signal pin (marked with a plus symbol) on the piezo speaker, and pin #2 (analog pin 1) is connected to the junction of the photocell and the resistor. Power and ground may be obtained from the USB connection or from an external supply connected to BAT+ (5 to 16 volts DC). Figure 3-15 shows a 5.5/2.1 mm power jack to wire adapter to use power from a common AC to DC transformer.

 You can connect a piezo speaker to pins #0, #1, and #2 with appropriate code changes. If it is connected to pins #3 or #4, it will interfere with the USB bootloader during uploads. If you decide to alter the circuit to use pin #3 or #4, disconnect the pin #3 or #4 wires while uploading.

Figure 3-16 shows a wired breadboard. Near the reset end of the Trinket is the photocell, with the 1,000-ohm resistor above that, near the top. The black circle is the piezo. The piezo element is polarized, so the positive lead should be connected to the Trinket and the negative to ground (GND).

Do not substitute a common speaker in place of the piezo without additional circuitry. A speaker has a low DC resistance and will try to draw too much current from the microcontroller.

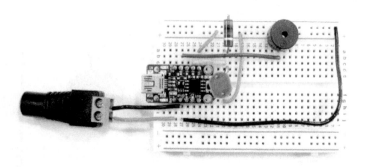

Figure 3-16. *The theremin on the breadboard*

Code

The sketch code is listed in Example 3-2. You can download the code from the repository for this book (*http://bit.ly/GettingStartedWithTrinket*), under the folder *Chapter 3 Code*, in the subdirectory *Chapter3_02Theremin*. How does it work? As more or less light hits the face of the CdS photocell, the resistance of the photocell changes. When the resistance changes, the voltage going into the analog pin (analog 1 is on the same pin as digital 2: pin #2) will vary up and down. This analog voltage is read by the analog pin. The combination of the resistor and the photoresistor is a form of *voltage divider*, a circuit that reduces the voltage between the two endpoints (in this instance, power and ground).

The voltage value read is *scaled*, or *mapped* (through the software `map` function that provides linear interpolation (*http://bit.ly/linear_interpola tion*)), to a sound frequency value. The piezo buzzer is then toggled on and off at the frequency calculated. The speaker output pin must be set as an output in the `setup` function using the `pinMode` function. Although the sound waveform will appear on the pin without it, it will not drive the piezo correctly.

Example 3-2. Sketch for the Trinket Theremin project

```
/* Trinket Theremin Sketch
Read the voltage from a cadmium sulfide (CdS) photocell voltage
divider and output a corresponding tone to a piezo buzzer.
*/

#define SPEAKER    0 // Speaker on Trinket pin #0
#define PHOTOCELL  1 // CdS photocell on Trinket pin #2 (analog pin 1)
#define SCALE      2 // You can change this to change the tone scale

void setup() {
  pinMode(SPEAKER,OUTPUT);    ❶
}

void loop() {
  int reading=analogRead(PHOTOCELL);    ❷

  int freq=220+(int)(reading*SCALE);    ❸
  beep(SPEAKER,freq,400);    ❹
  delay(50);    ❺
}

// The sound-producing function    ❻
void beep (unsigned char speakerPin,
           int frequencyInHertz,
           long timeInMilliseconds) {
  int x;
  long delayAmount = (long)(1000000/frequencyInHertz);
  long loopTime = (long)((timeInMilliseconds*1000)/(delayAmount*2));
  for (x=0; x<loopTime; x++) {
    digitalWrite(speakerPin,HIGH);
    delayMicroseconds(delayAmount);
    digitalWrite(speakerPin,LOW);
    delayMicroseconds(delayAmount);
  }
}
```

❶   It is important to configure the piezo pin for digital output. If you don't do this, you won't hear anything. If you were to connect an oscilloscope to the pin, you'd see the pin change frequency but current would not flow to the piezo.

❷   The voltage proportional to the light hitting the photocell is read on analog pin 1, which is Trinket pin #2 (see the diagram in Chapter 1).

❸   The analog value is scaled to start at 220 Hz and rise based on the voltage multiplied by a value in SCALE that the programmer may change as desired in the code.

❹   The Arduino **tone** function does not work for the ATtiny85 on the Trinket. The **beep** function is similar and will work for any Trinket data pin.

The duration is set to 400 milliseconds; this may be changed in the code also.

❺ The delay value (in milliseconds) also is not critical and may be changed.

❻ The beep code is similar to that from Dr. Leah Buechley (*http://web.media.mit.edu/~leah/LilyPad/07_sound_code.html*).

A less math-intense function to produce sounds is introduced in the Trinket Animal project in Chapter 6.

Use

When the circuit is powered up, it should emit a sound. If it doesn't, unplug the power and check the connections.

A tone at a single frequency might start to annoy folks. Cup your hand and start to block the light hitting the photocell. You will find the tone changes! Now you can "play music" by varying the amount of light the photocell receives at any given time.

You can change the SCALE variable in the program or even the entire calculation of the frequency freq to get different sound ranges. You may vary the frequency calculation, the length of time the tone is generated, and the delay between tones.

Sound and Music

Musical notes are very specific types of sound vibrations. These vibrations are measured in *frequency*, or how fast the sound wave vibrates per second. Frequency is measured in units of Hertz (cycles per second, abbreviated Hz). You can generate precise musical notes by selecting the correct frequency.

The musical scale is in octaves of seven whole notes. If you use half steps (sharps, which are followed by a # symbol, and flats), there are 12 notes. Musical frequencies are listed in Table 3-1.

Table 3-1. *Frequencies associated with musical notes*

Note	Frequency (Hertz)
A_2	110.00
A#	116.54
B	123.47
C	130.81
C#	138.59
D	146.81
D#	155.56

Note	Frequency (Hertz)
E	164.81
F	174.61
F#	185.00
G	196.00
G#	207.65
A_3	220.00

There is a mathematical relationship in the numbers. More information on the math of music can be found on Wikipedia (*http://en.wikipedia.org/wiki/Mathematics_of_musical_scales*).

Let's reexamine the Theremin program code. The frequency of the tone played was calculated by `freq=220+(int)(reading*SCALE);`.

The minimum frequency you can generate with this piezo is 220 Hz, which is the musical note A. If the reading on the photocell is zero, then the frequency will be an A. If the reading is higher, the calculated frequency increases, which makes the notes higher. The value of `SCALE` is introduced to vary how much each change in the reading adds to the frequency. The `analogRead` function outputs from 0 to 1023. The `SCALE` in the code is set to 2, so the maximum frequency possible would be 220 + (1023 * 2) = 2266 Hz, which is just above $D\sharp_7$ (rather high). You can see that the calculation will not map to exact musical notes. Changing `SCALE` would raise or lower the range of notes played. Perhaps true musicians will cringe at a performance?

The piezo speaker does not have a wide frequency response (the span of frequencies it can accurately produce). A typical paper cone speaker has a better audio frequency response. However, speakers require an audio amplifier to be added to a circuit. This boosts the signal and makes the input level compatible with the output signal (this is called *impedance matching* and *driving*).

Conclusion

This chapter progressed from connection basics to actual programming. In the next chapter, I'll discuss libraries in depth, letting you create even more complex and fun programs.

4/Libraries and Optimization

Libraries are code written by others to provide specific software functionality that you can use in your own sketches. The Arduino community has hundreds of prewritten libraries, most of which are free. This wealth of code allows hobbyists and professionals alike more time to focus on their intended projects, and not the specifics of a specialized chip or algorithm.

Many of the libraries currently available are written for specific Atmel or other companies' processors and may not work with other microcontrollers. Because it uses the ATtiny85 chip, with its different memory and architecture, the Trinket has some differences from the Arduino Uno and other popular boards.

This chapter introduces you to many Trinket-compatible libraries and provides pointers to places where more are being developed.

Arduino Libraries

You can start to browse the libraries that are bundled with the Arduino IDE by clicking on File, then Examples. All of the entries below ArduinoISP are library examples (see Figure 4-1).

The ATtiny processors, to provide several types of communication protocols with a small number of pins, combine hardware functionality among pins. ATtiny maker Atmel calls its scheme *Universal Serial Interface* (USI). Standard Arduino libraries for communications (serial, Two-Wire/I^2C, and Serial Peripheral Interface/SPI) will not work with the USI architecture. Fortunately, creative folks have written alternate libraries that give us nearly the same functionality. This may require you, however, to make changes to how you would normally write Arduino code to run on a Trinket.

Figure 4-1. *The Arduino IDE built-in libraries list*

The issues you may encounter when selecting Arduino libraries for use on a Trinket include:

1. Differences between the USI hardware and serial, I²C, and SPI libraries.

2. The two timer circuits on the ATtiny85 are 8-bit, instead of the multiple 16-bit timers on the Uno. Libraries that use timers should be reviewed for compatibility. One significant issue is that the standard Servo library will not work, but there are Trinket-compatible replacements.

3. Memory limitations. A program or a library that uses a large amount of RAM, library, or both can exceed the Trinket's onboard memory.

4. Floating-point (decimal number) math functions take a great deal of code space—up to 2,000 bytes. If you can limit your code to integer (whole) numbers, you can save memory.

Fortunately, there are ATtiny-optimized libraries for many functions.

ATtiny-Optimized Libraries

You can find libraries optimized for the ATtiny85 in several Internet repositories. Most Adafruit libraries are available on the code-sharing site GitHub (*https://github.com/adafruit/*).

The following standard Arduino IDE libraries work well with the Trinket:

SoftwareSerial
>The library included with the Arduino IDE. This works with the Trinket as is. You can use any of the five data pins for serial communication; just be careful of the resistors on certain pins, as noted in Chapter 3.

EEPROM.h
>The Arduino EEPROM persistent memory library. For an example of its use with the Trinket, see *http://bit.ly/secret_knock_drawer*.

avr/power.h
>Controls some of the ATtiny85 power functions.

avr/sleep.h
>Controls putting the chip into a sleep mode.

Other standard libraries may also work, but not all have been tested by the Trinket community.

The following third-party libraries have been tested and work with Trinket:

SendOnlySoftwareSerial
>This library by Nick Gammon is discussed and linked to in the Arduino forum post at *http://bit.ly/SoftwareSerial_topic*.

Adafruit_SoftServo (http://bit.ly/SoftServo)
>Similar to the Arduino Servo library, with a tutorial (*http://bit.ly/Servo_Control*).

TinyWireM
>This is an implementation of the Wire/I^2C library (Master mode) that uses the USI in the ATtiny85. For Trinket, Adafruit has a fork of the code (*https://github.com/adafruit/TinyWireM*) that works very well. Any library that uses the Arduino Wire library may need to be modified to refer to TinyWireM.

TinyWireS (https://github.com/rambo/TinyWire)
>The Wire library (slave mode).

TinyLiquidCrystal (https://github.com/adafruit/TinyLiquidCrystal)
This is used to drive liquid crystal displays, including those controlled by the MCP23008 driver chips.

Adafruit_NeoPixel (https://github.com/adafruit/Adafruit_NeoPixel)
The standard Adafruit smart RGB LED (WS2812B) driver library is Trinket compatible. See also Adafruit's extensive guide to NeoPixels (*http://bit.ly/NeoPixel_uberguide*).

FastLED (https://github.com/FastLED/FastLED)
An awesomely optimized library for driving many types of smart LED products and more. Trinket and Gemma support is included.

Adafruit_LEDBackpack (http://bit.ly/LEDBackPack_Library)
Controls Adafruit LED displays. This library requires the Adafruit_GFX library even if your project does not render graphics.

Adafruit_GFX (http://bit.ly/Adafruit_GFX)
The Adafruit graphics library works with Trinket, although it takes a great deal of memory. It is used in the Trinket Occupancy Display project in Chapter 6, but fits with only a few bytes to spare.

TinyAdafruit_RGBLCDShield (http://bit.ly/RGB_Shield)
Allows control of the Adafruit RGB LCD Shield normally used with larger Arduinos.

TinyDHT (https://github.com/adafruit/TinyDHT)
Library compatible with DHT11/21/22 temperature and humidity sensors using integer math.

TrinketKeyboard (http://bit.ly/Trinket_USB)
Emulates a keyboard through the Trinket USB port (which is tricky, as the port is not a standard USB port)—see *http://learn.adafruit.com/trinket-usb-keyboard* for use.

TrinketHidCombo (http://bit.ly/Trinket_USB)
Provides human interface device (HID) functions, such as mouse or keyboard emulation, via the USB port. See *http://bit.ly/Volume_Knob* for an example of using this library.

TinyRTClib (https://github.com/adafruit/TinyRTCLib)
Library for the Adafruit DS1307 real-time clock module.

Adafruit_TinyFlash (http://bit.ly/TinyFlash)
Allows read/write from Winbond flash chips via SPI.

TinyNarcoleptic
A fork of the Google library for putting the processor into a low-power state. See *http://bit.ly/Narcoleptic_lib_topic* for information. For other sleep modes for the ATtiny85, see *http://bit.ly/ATtiny85_Sleep_Modes*.

Servo8Bit
>Some implementations of this library fail to work on the Trinket. The Adafruit servo library listed earlier is recommended. The version at *https://github.com/solderspot/Servo8Bit* is reported to work by an Adafruit forum user.

tinySPI (https://github.com/JChristensen/tinySPI)
>This library by Jack Christensen is an Arduino SPI master library for ATtiny44/84/45/85 that utilizes the USI hardware in the ATtiny microcontrollers.

Arduino-UsiSerial (http://bit.ly/UsiSerial)
>This library by Frank Zhao provides hardware serial control via the USI interface on Trinket pins #0 and #1 with a default baud rate of 19,200.

VirtualWire
>A library enabling use of amplitude shift keying (ASK) radio transmission on inexpensive radio frequency (RF) transmitters. An ATtiny85-compatible version is available at *http://bit.ly/VirtualWire*.

TinyNewPing (http://bit.ly/TinyNewPing)
>This library by an engineering student named Matthew allows use of common ultrasonic sensors such as the HC-SR04 on the ATtiny85.

arduino-nrf24l01 (http://bit.ly/nrf24l01_library)
>This library by Abe Connelly, combined with a small circuit, allows the use of a nRF24L01 radio module on an ATtiny85 using only three pins (rather than five).

New libraries are written often by enterprising individuals. Use an Internet search engine to search for terms "ATtiny85" or "Trinket" along with a the term describing the functionality you are looking for. Some of the projects you may find are written for other programming environments or ATtiny85 boards, so you may need to modify the code. Internet search results could be a starting point in building the code you want to use.

The Adafruit Learning System (*http://learn.adafruit.com/*) has new Trinket and Gemma tutorials added periodically.

See Chapter 7 for additional resources for Trinket information online.

Installing Libraries

To use a library, you must obtain the code and place it where the Arduino IDE can find it for use in programs.

Most websites hosting code that is identified as a library, including Adafruit and GitHub, will offer downloads of the library file as a zipped archive. You can generally open the ZIP file with Windows, Mac, or Linux and find the files needed inside. You will want to copy the files out of the ZIP file, but where should they go?

Where to Install Libraries

It is important to place your library code files in the correct location. Otherwise, the Arduino IDE will not be able to locate them when you try to compile and upload your sketches.

Locate your sketchbook folder. Your sketchbook folder is the folder where the Arduino IDE stores your sketches. This folder is automatically created by the IDE when you install it.

On Windows and Macintosh machines, the default name of the folder is *Arduino* and the default location is in your *Documents* folder (see Figure 4-2).

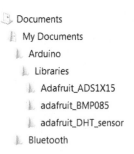

```
Documents
   My Documents
      Arduino
         Libraries
            Adafruit_ADS1X15
            adafruit_BMP085
            adafruit_DHT_sensor
         Bluetooth
```

Figure 4-2. *The subdirectories where libraries are located*

On Linux machines, the folder is named *Sketchbook* and it is typically located in */home/<username>*.

This is the only difference between libraries on Linux versus Windows and Mac machines: your sketchbook folder is named *Sketchbook*, not *Arduino*.

User-installed libraries should go in a folder under your sketchbook folder (a subfolder) named *Libraries*. This is where the IDE will look for user-installed libraries. To locate your sketchbook folder, open the Preferences dialog box, as shown in Figure 4-3, by clicking File→Preferences in the IDE. The path to this folder is given in the "Sketchbook location" field at the top of the Preferences dialog, as seen in Figure 4-4.

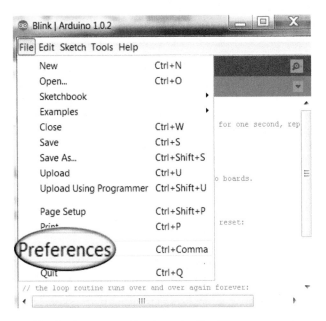

Figure 4-3. *The Preferences option is under the File menu*

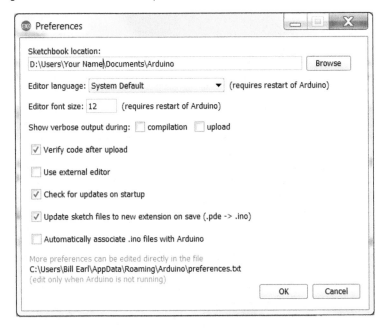

Figure 4-4. *The Preferences dialog box*

Once you know the location, navigate to this folder in File Explorer (Windows), Finder (Mac), or the shell (Linux). Now you're ready to install the library:

1. First make sure that all instances of the Arduino IDE are closed. The IDE scans for available libraries only when the program starts up. It will not see new libraries as long as any instance of the IDE is still open.

2. Download the ZIP file. Click the Download ZIP button on the GitHub repository page (Figure 4-5). Alternatively, if the code is not in GitHub, download it from the Internet site hosting the library.

3. Follow the steps listed in the following sections.

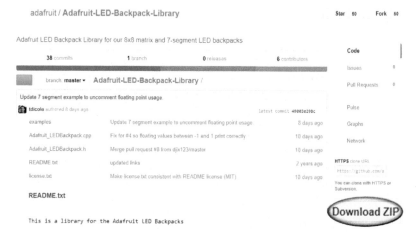

Figure 4-5. *The Download ZIP button on GitHub*

Installing a Library in Windows

Follow these steps to install a library in Windows.

1. Open the ZIP file and copy the library's top-level folder or whatever the main folder is named, as shown in Figure 4-6.

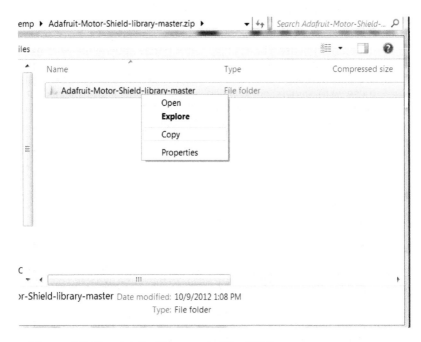

Figure 4-6. *Copying the library out of the ZIP file*

2. Open your sketchbook *Libraries* folder and paste the folder you copied from the ZIP file into it (see Figure 4-7). If the *Libraries* folder doesn't exist, create it.

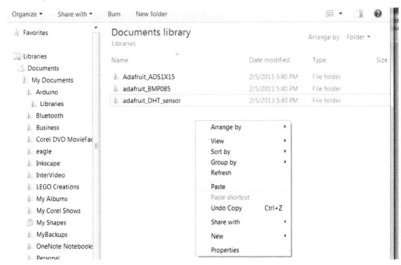

Figure 4-7. *Pasting the library in your Arduino libraries folder*

3. Give the library folder a valid name. The name of the folder you've copied may be different from the name the Arduino IDE expects to find during normal use. For example, GitHub may append the word *master* to the filename, which is not desired in the final installation. Also, the IDE will not recognize folders with dashes in the name (you can replace these with underscores, which are fine). Check the main code or library examples for the name that the example code expects, and change the name to that, as shown in Figure 4-8.

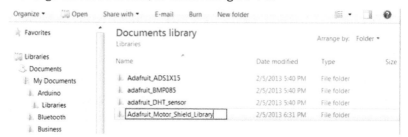

Figure 4-8. *Rename the library folder to the name expected by your program*

4. Restart the Arduino IDE and verify that the library appears in the File→Examples menu (Figure 4-9). Load one of the library examples to test, if there is one (note that some examples may require other libraries).

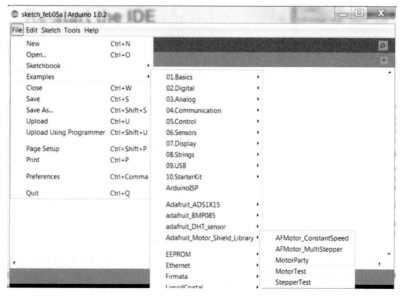

Figure 4-9. *When you restart the IDE, your library should be listed*

5. Verify programs that use the library compile. Ensure you have all the needed libraries installed. Click the checkmark icon in the upper left (compile and upload) and verify that an example sketch (or whatever program you have) compiles without errors. If many errors are generated, compare the library folder name against what the Arduino IDE believes the name of the library should be. If there is a difference, rename the library folder.

Installing a Library in OS X

The following steps allow you to install a library in Mac OS X:

1. Find the downloaded library in the *Downloads* folder. OS X will usually save the ZIP file in this folder by default (see Figure 4-10). If your browser doesn't automatically open the ZIP file, double-click it to extract it.

Figure 4-10. *Finding the downloaded library*

2. Open your sketchbook *Libraries* folder and drag the main folder from *Downloads* into it (Figure 4-11). If the *Libraries* folder does not exist, create it.

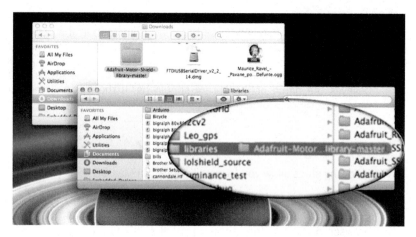

Figure 4-11. *Drag your library into the Arduino Libraries folder*

3. Give it a legal name. The library, if downloaded from GitHub, will have the word "master" appended to its name. Remove that from the name prior to use. Also, the IDE will not recognize folders with dashes in the name. You can replace these with underscores, as shown in Figure 4-12.

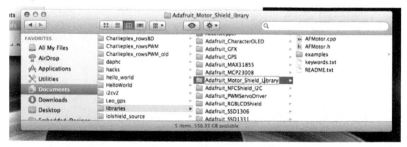

Figure 4-12. *Rename the library if necessary*

4. Restart the Arduino IDE and verify that the library appears in the File→Examples menu (Figure 4-13).

If the library has example programs included, testing whether an example compiles in the IDE is a great way to be sure the library is functional. If the example works, you can proceed to test your own code that uses the library. If code that uses the library does not compile, see Chapter 8 to troubleshoot the library installation. Bear in mind that some programs rely on other libraries, which will also need to be installed.

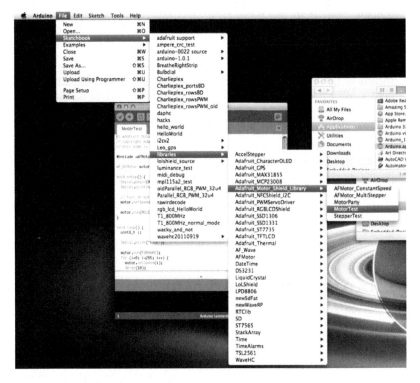

Figure 4-13. *Viewing the libraries in the Arduino IDE*

Using Libraries

To use a library, use the #include statement near the top of the sketch. For example, to use the SoftwareSerial library, place the following line at the top:

```
#include <SoftwareSerial.h>
```

The .h at the end of the name SoftwareSerial tells the processor this is a *header file*, which defines objects and code for use later in the program. See the examples throughout the book for how libraries are included.

At times, you may want to use several libraries in one program. In some cases, libraries require other libraries: any library performing I²C communication will need the TinyWireM library, for example, and some Adafruit display libraries require the Adafruit_GFX library.

Sometimes you will just wish to add two or more libraries for their functionality. But as you layer on such code, the amount of program space the libraries require increases. Using several large libraries probably will require more than the 5,130 bytes of flash program memory available (recall that the bootloader takes up the rest of the 8,192 bytes of program memory on the ATtiny85).

When two libraries want to use the same ATtiny85 resource (one that is not meant to be shared), they can conflict. For example, the SoftwareSerial library will not work with the TrinketKeyboard library, due to both wanting the same change interrupt vector (an ATtiny85 hardware resource). It may or may not be possible to edit a library to avoid a conflict.

You should carefully decide which libraries are required for your project. Try using libraries one at a time and see how much program space is left for additional functionality. If adding a library exceeds the available space, remove the #include line and decide what to do next. If your project cannot do without the affected code, you might have to consider another micro-controller platform. This is discussed in Chapter 7.

Some code use could bloat your program without your knowledge. Certain optimizations may help, which I discuss next.

Library Issues and Limitations

If you get Arduino IDE errors when using a library, see "Common Library Problems" on page 217 for common issues.

Libraries written by third parties may not take into account the resources or situations encountered by other users. Although libraries provided by experienced developers may be better tested, they may not be ideal for your application. Most carry open source licenses, but some do not. And many hobbyist libraries may not be crafted to the same level as the examples in the Arduino IDE. If you are wondering what you are getting yourself into, this is good. But not to worry! In most cases you should read what the library is intended for, study the documentation or examples, including function calls and parameters, then try it. It may not be exactly what you want; if not, you can delete the library and move on. If the library provides the functionality you like, however, you have more time to work on your project.

Memory Optimizations

Getting the most out of the Trinket without exceeding its capabilities may require some optimization. Some programmers use coding methods on the Arduino Uno and Mega that are not optimized but work well due to the larger resources on those controllers. But for the Trinket, practices may harken back to the 1980s, when every byte of memory was to be conserved. Be it program space or RAM/variable space, the methods discussed here can be used to reduce the amounts required for running your code.

Program Space Optimization

Here are some tips for reducing your program size:

- Avoid floating-point (decimal) numbers. The full floating-point library can be 2,000 bytes or more. That is three-fifths of your available program space. If you can use integer (whole) numbers only, you will save considerable space. The TinyDHT library mentioned earlier is an integer-optimized version of the full Adafruit DHT library that you can use to save space.

- Don't use math functions. Using pow(x,2) to get x^2 could incur more program space than x*x. Likewise, the sqrt (square root) and trigonometric functions can be very large. Other language functions may be much smaller. Experiment to see which ones create issues.

- Comment out code you will not use. If you have code your program will not use, enclose it within a /* ... */ comment block so the compiler will ignore it. This includes any functions you do not plan to call.

- Reuse code. If a portion is used in different parts of a program, consider placing that code in a function and call the function when needed.

Variable Optimization

Most users do not declare large amounts of variables, but it is possible that you may call code that does. The Adafruit-NeoPixel library dynamically allocates memory for each pixel to store the red, green, and blue numeric colors. Approximately 110 pixels can be used without other variable usage, often less with your own use. Displays often use RAM if the display hardware does not provide its own display storage (called *buffering*).

You can check the amount of remaining memory with the code shown in Example 4-1.

Example 4-1. Function to return the amount of free RAM on an Arduino compatible

```
int freeRam () {
  extern int __heap_start, *__brkval;
  int v;
  return (int) &v - (__brkval == 0 ? (int) &__heap_start : (int) __brkval);
}
```

Typical use might be if(freeRam() < 100) digitalWrite(1,HIGH);, which would turn on the pin #1 red LED if RAM is low.

The following are ways to save valuable RAM:

- Do not declare unneeded variables. The compiler most likely will detect this and save you, but then again, it may not.

- Do not include unneeded libraries.

- If a variable is only needed in the scope of a function, declare it in the function and not globally. The dynamic memory allocator for functions will clean these up if they're local. Pass variables as function arguments, such as `myfunction(variable1, variable2)`, rather than using global variables if possible.

- If you have fixed values that never change, declare them static, as in `static int MinutesPerHour = 60;`. The compiler should optimize these values.

You can also control the size of various signed and unsigned integers. The C language is not explicit when you declare a variable as type integer or unsigned integer, but the Trinket works most effectively with 8-bit data (as the ATtiny85 is an 8-bit microcontroller). You can be sure to get an 8-bit integer when desired by declaring variables as `int8_t` or `uint8_t` for signed and unsigned integers, respectively. The values that can be represented by 8 bits are:

- 8-bit signed integer(`int8_t`): −128 to 127
- 8-bit unsigned integer(`uint8_t`): 0 to 255

For larger numbers, you can use 16-bit integers:

- 16-bit signed integer (`int16_t`): −32,768 to 32,767
- 16-bit unsigned integer (`uint16_t`): 0 to 65,535

For even larger numbers, you can use 32-bit numbers:

- 32-bit signed integer (`int32_t`): −2147483648 to 2147483647
- 32-bit unsigned integer (`uint32_t`): 0 to 4,294,967,295

If your numbers will not exceed certain ranges, you can optimize your code and memory usage by using the correct declarations. The caution is that if a value does exceed a range, an error will not be generated and the wrong value will be registered. Consider the following statements:

```
uint8_t var = 255;
var = var + 1;
```

The value of **var** will not be 256! It will wrap around via binary math to zero. Check your program, and if a value will grow, allocate enough space for the maximum value. If only positive numbers will be used somewhere, you can consider an unsigned integer. For example:

```
for (i=0; i<100; i++) { j = i + 100; }
```

We can declare variable i an int8_t or a uint8_t. j will exceed 127, so it should be at least a uint8_t, or maybe a 16-bit number if additional math may make it go higher than 255.

Saving one or two bytes in an entire program may not make much of a difference overall, but it could if memory is short. Also, if you declare many variables, these small savings could make all the difference. This can be especially true when using arrays. For example, int16_t array[100]; takes 200 bytes of RAM, while int8_t array[100]; takes half as much.

More exotic methods are used by some folks to reduce memory use, but judicious variable usage and good coding practices are highly recommended to save memory.

Conclusion

Libraries can greatly expand program functionality while giving you more time to focus on designing and building your project. Not all Arduino libraries work with the Trinket, due to the ATtiny85 processor's limitations. There are still many libraries that do work, though, and optimizing code and memory usage will provide the ability to fit projects into the constraints the Trinket presents. Although some believe the petite amount of memory the Trinket offers is problematic, most recognize it as a challenge to build projects to fit the Trinket's capabilities. It is a slightly more refined process, with the reward being amazing power in such a small package.

5/Intermediate Projects

Now that you know how to harness the power of libraries in your programs, you can build many additional projects that demonstrate the versatility of the Trinket. Starting with the popular NeoPixel LEDs, the projects in this chapter use servos, sensors, serial communication, and real-time clocks. The principles in these projects translate over the wide variety of technologies used in Maker projects.

These projects have build details that require more advanced skills or tools. Younger Makers may need assistance from more experienced builders.

Controlling Smart LEDs: NeoPixels

The way we use single-color LEDs (introduced back in Chapter 3) has not changed much in the last 30 years. About 15 years ago, tricolor LEDs in a single package appeared. Containing a red, green, and blue LED in one package, these LEDs were placed into art and the first large color displays. They are still widely used today, but they have some characteristics that make them hard to use in small projects. They have three control lines and one common lead. Our five-pin Trinket would need to give up three valuable data lines to drive a single tricolor LED.

Fortunately, innovation in LED technology continues. In recent years, LEDs with built-in control chips have revolutionized how we connect and use LEDs. Small flat-package LEDs only require connections to power, ground, and a single digital signal line to set the intensities of the red, green, and blue LEDs in creating the desired color. Adafruit's NeoPixel (*http://www.adafruit.com/neopixel*) (see Figure 5-1) is a product line of such LEDs, packaged conveniently in breadboard-friendly versions, rings, strips, and sewable versions.

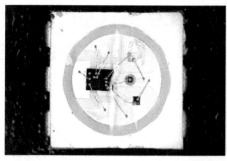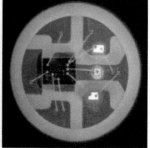

Figure 5-1. *Close-ups of a WS2812B smart LED (Adafruit's NeoPixel)*

The innovation in LED connections has also given us the power to chain multiple LEDs together without additional data pin connections to a Trinket. An output pin on one NeoPixel connects to the input pin on the next. The number of NeoPixels is limited only by the amount of RAM available on the microcontroller (to hold the color values). The limit has been tested at approximately 110 pixels for the Trinket using Adafruit's NeoPixel library. This equates to 330 bytes of memory, leaving 182 bytes set aside by the compiler for variables and the stack (function value storage). Even with a large number of pixels, only one Trinket data pin is needed to control an entire string.

Important Things to Know About NeoPixels

Not all addressable LEDs are NeoPixels. "NeoPixel" is the Adafruit brand for individually addressable RGB color pixels and strips based currently on the WS2812B LED/drivers, using a single-wire control protocol. Other LED products—WS2801 pixels, LPD8806 strips, and "analog" LED strips—use different control methodologies (and have their own interface methods). If you want to build a project that specifies NeoPixels, be sure you obtain the correct parts. If you have one of the other types of smart LED products, see the FastLED library listed in Chapter 4.

Unlike a regular LED, a NeoPixel does not just light up when power is applied; it requires a microcontroller (such as a Trinket) and some programming to send specific control information to its data pin. This programmability allows your code to create effects and animations.

Each NeoPixel has a data in (Din) digital signal line and a data out (Dout) line. To control multiple pixels, you only have to connect the Dout pin of a NeoPixel product to the Din of the next product, and so on down the line. You start the numbering from the first pixel (number zero) to the last through the connections you make.

NeoPixels are not the answer for every project, but their flexibility and packaging make for beautiful and compelling displays.

NeoPixel Packaging

NeoPixels come in a variety of layouts: single pixels, strips, rings, sticks, and matrices (see Figure 5-2). New products come out periodically, as this is a fast-growing hobbyist area.

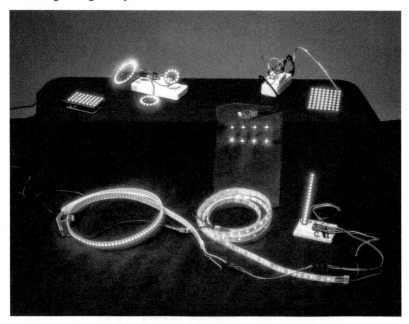

Figure 5-2. *A variety of individual NeoPixels along with strips, rings, and matrices*

NeoPixel Ornaments

One of the most popular uses of NeoPixels is creating simply amazing decorations. With a simple circuit, you can create a variety of colors and patterns.

Figure 5-3 shows the diagram for a simple two-pixel circuit powered over the USB port. USB 2 can supply 500 milliamperes (mA), or 0.5 amperes. Each pixel may draw a maximum of 60 mA. Powering a Trinket and two pixels is well under the USB current limit. In later projects, we'll use an external power source to allow for using many more pixels.

Parts List

- Trinket 5V, Adafruit #1501 or Maker Shed #MKAD69
- Breadboard-friendly NeoPixels (4-pack), Adafruit #1312 or Maker Shed #MKAD60
- Half breadboard, Adafruit #64, Maker Shed #MKKN2, or equivalent (a full breadboard is fine)
- Breadboard wires, Adafruit #153, Maker Shed #MKSEEED3, or equivalent
- Optional: 4xAA battery holder, Adafruit #830, and 4 AA batteries

Build

We'll take power from the Trinket USB+ pin for this project. The NeoPixel *drive signal* is from Trinket pin #0. You wire the two NeoPixels in series, as shown in Figure 5-3—the pin #0 signal goes to the first NeoPixel Din pin, and the first NeoPixel Dout (labeled O) connects to the second NeoPixel Din pin.

Made with **Fritzing.org**

Figure 5-3. *Controlling two NeoPixels with Trinket*

If you plan to power this type of circuit with batteries, connect the positive terminal of a 4 AA battery pack to the Trinket BAT+ pin and the black wire to the GND pin. The Adafruit AA holder has an on/off switch for convenience.

Next, upload the program shown in Example 5-1 from the repository for this book (*http://bit.ly/GettingStartedWithTrinket*) (directory *Chapter 5 Code*, subdirectory *Chapter5_01TwoNeoPixels*).

Example 5-1. Testing Two NeoPixels with Trinket

```
/* Getting Started with Adafruit Trinket - Two NeoPixel Program */

#include <Adafruit_NeoPixel.h> // Add in the Adafruit-NeoPixel library    ❶

#define PIN       0 // NeoPixel signal output pin    ❷
#define NUMPIXELS 2 // How many pixels we will use

// Set the NeoPixel data structures and operation

Adafruit_NeoPixel strip =
  Adafruit_NeoPixel(NUMPIXELS, PIN, NEO_GRB + NEO_KHZ800);

void setup() {    ❸
  strip.begin(); // This sets up NeoPixels for use
  strip.show();  // Initialize all pixels to 'off'
}

void loop() {
  int8_t i;    ❹

  for(i=0; i <= 255; i++) {    ❺
    // Cycle the first NeoPixel up in blue, the second down in red
    strip.setPixelColor(0,strip.Color(   0,0,i));    ❻
    strip.setPixelColor(1,strip.Color(255-i,0,0));
    strip.show();
    delay(6);
  }

  delay(100);

  for(i=255; i >=0 ; i--) {
    // Cycle the first NeoPixel down in blue, the second up in red
    strip.setPixelColor(0,strip.Color(   0,0,i));
    strip.setPixelColor(1,strip.Color(255-i,0,0));
    strip.show();
    delay(6);
  }

  delay(100);
}
```

❶ You must include the Adafruit-NeoPixel library, which you can obtain from *https://github.com/adafruit/Adafruit_NeoPixel* and install per the instructions in Chapter 4.

❷ Here, you make some definitions allowing you to easily change things: the pin on the Trinket to connect to the NeoPixels and how many pixels to drive. The last values—NEO_GRB and NEO_KHZ800—are the values for all the NeoPixels now sold by Adafruit (other values can be used with some other brands of smart LEDs).

❸ The setup function initializes the data structure for the NeoPixel strip. The loop cycles the color intensity of the LEDs; the first one cycles up then down in blue and the second cycles down then up in red.

❹ Variable i is an 8-bit unsigned integer (uint8_t), which is big enough as long as you restrict its use to between 0 and 255, as described in "Variable Optimization" on page 51.

❺ Each time through the loop function, you directly manipulate the blue value (the third parameter in the Color function) and the red value (the first parameter). The green value (the second value) is left at zero. If you prefer green over red, you can change it to a fixed number or cycle it as a function of variable i or 255-i.

❻ The strip.Color function takes three numbers, a red, a green, and a blue value, from 0 to 255 (low/off to high/full on). The return value is a single 32-bit (long) integer for the combined color. The strip.set-PixelColor call takes two values: the pixel number and the 32-bit color number. This is why we have strip.Color nested into the set-PixelColor function call.

Be sure you have selected Trinket 8 MHz in the Arduino IDE Tools→Board menu. If this setting does not match the code used, the NeoPixels may not receive the correct timing signals and appear as all white LEDs, all off, or exhibit other unexpected behavior.

With these basic connections and some programming, you can build a wide variety of projects. Single pixels are useful on very small projects such as ornaments. You can bundle a Trinket, NeoPixel(s), and a battery into any semitransparent enclosure for a wonderful display (Figure 5-4). If you have a 3D printer, you can create complex custom enclosures for your NeoPixels in a variety of colored plastics. A craft store provides a range of materials that can be used for ornamental projects.

Alternatively, you can use a Trinket and two NeoPixels to make cute blinky-eyed toys. One great project is an interactive toy (*http://www.manoelle mos.com/interactive-toy-with-adafruit-trinket-and-neopixels/*) by Manoel Lemos (Figure 5-5).

Figure 5-4. *A NeoPixel tree topper ornament by Rick Winscot*

Figure 5-5. *An interactive Trinket and NeoPixel toy by Manoel Lemos*

LED Color Organ

Once you have multicolored lights, you can use sensors to have them respond to the world around us. A popular electronics project involves

LEDs flashing to sound, such as you might see at a concert. In the 1970s, incandescent color organs were popular, with their lights flashing to music (usually disco). This project is a Trinket circuit that provides this same effect for your next concert.

How It Works

Color organs sample sound and flash lights based on either the sound intensity or frequency. The higher-end units use analog or digital signal analysis to determine the sound energy in selective parts of the frequency spectrum and flash the lights accordingly.

The Trinket can provide sound intensity sampling and display any number of colored patterns based on music volume, as shown in Figure 5-6.

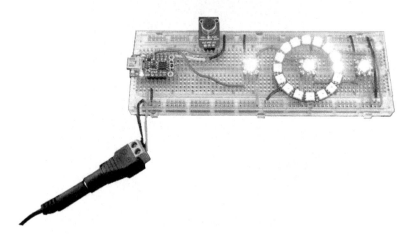

Figure 5-6. *The NeoPixel color organ on a breadboard*

Parts List

You have the freedom to choose an array of NeoPixel types to make the project. You may want to sketch out your own light layout, too.

- Trinket 5V, Adafruit #1501, or Maker Shed #MKAD69
- Electret microphone amplifier MAX4466, Adafruit #1063
- NeoPixels: Flora RGB V2 pack of 4, Adafruit #1260, or breadboard-friendly RGB NeoPixels, Adafruit #1312 or Maker Shed #MKAD60
- NeoPixel ring(s): 16-pixel Adafruit #1463 or Maker Shed #MKAD75, 24-pixel Adafruit #1586, 12-pixel Adafruit #1643
- 10,000-ohm (10K) potentiometer, Adafruit #562 or similar

- 5V, 2A power supply, Adafruit #276 or similar
- Female 2.1/5.5 mm DC power connector, Adafruit #368
- Full breadboard, Adafruit #239, Maker Shed #MKEL3, or similar
- Hookup wire, Adafruit #289, #288, #290, or similar (such as Maker Shed #MKEE3)
- Optional: Full perma-proto board for permanent build, Adafruit #1606 or Maker Shed #MKAD50
- Optional: Other NeoPixel products
- Optional: 5V, 10A power supply (to power many, many pixels), Adafruit #658

Build

The best way to start is to breadboard first, as shown in Figure 5-7. You can then transfer the circuit to a small perma-proto circuit board when you are satisfied with your circuit and want to consider a permanently mounted project.

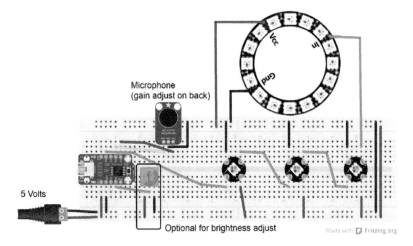

Figure 5-7. *Wiring diagram for the color organ*

If you haven't already, solder the headers supplied to the Trinket's pins to facilitate breadboarding, as directed in "Preparing the Trinket" on page 17. You can also solder a small three-pin header on the microphone breakout board for a breadboard connection. For a more permanent circuit, you could use a servo extension cable to extend the microphone. Alternatively, you can solder three wires from the microphone breakout to the Trinket, power, and ground lines to extend placement of the microphone through an enclosure.

You can use a range of NeoPixels for your project, up to a limit of about 110 (with appropriate power supply). The single NeoPixels pictured have single header pins soldered on. The ring was placed over the middle single LED. If you plan to build a permanent circuit, wiring to the pads would be a better choice.

 NeoPixel Connections

Connect the Dout pin of one NeoPixel product to the Din pin of the next. When programming, the first NeoPixel (numbered zero) is the first one in the chain, with each one following the wiring to the last pixel on the last product.

A good 5-volt power supply is very important if powering more than three NeoPixels. Calculate the maximum power by multiplying the number of pixels times 0.060 amperes, then add 0.050 amps for the Trinket circuit; that is the minimum current you should budget for. Select a power supply that can provide more than your anticipated maximum current. For large, power-hungry strips, the Adafruit 5-volt, 4-amp, or 10-amp supplies or equivalent could be required.

Power for the microphone breakout is taken from the Trinket 5-volt regulated power pin.

Trinket pin #2 is both an analog and a digital pin. This circuit uses it as analog pin 1 to read the varying voltage from the microphone breakout.

Trinket pin #0 is used as the digital signal line out to the string of NeoPixels.

If you would like to control the brightness of the pixels, you can add a potentiometer (nominal 10 kiloohms; anywhere between 1,000 ohms and 1 megaohm should be fine). The center wiper is connected to pin #3 (analog 3 in the Arduino IDE). The changing voltage is read and mapped to the range the NeoPixel setBrightness function uses to set the pixel brightness. If you're not using the potentiometer, comment out the value POT_PIN by prefixing that line of code with //, and the code will use the maximum brightness.

If you have trouble loading programs after adding the potentiometer, this is due to pin #3 being shared with the USB port. Temporarily remove the Trinket from the circuit to load the program, then place it back into the circuit. If you make a permanent circuit board, use female headers as a socket to plug your Trinket into for ease of programming.

The code for this project is in Example 5-2 and can be downloaded from the repository for this book (*http://bit.ly/GettingStartedWithTrinket*) (directory *Chapter 5 Code*, subdirectory *Chapter5_03Organ*).

Example 5-2. The LED Color Organ code

```
/* LED "Color Organ" for Adafruit Trinket and NeoPixel LEDs*/

#include <Adafruit_NeoPixel.h>

#define N_PIXELS   19  // Number of pixels you are using      ❶
#define MIC_PIN     1  // Microphone is attached to Trinket Pin #2/(A1)   ❷
#define LED_PIN     0  // NeoPixel LED strand is connected to Pin #0   ❸
#define DC_OFFSET   0  // DC offset in mic signal - if unsure, leave 0
#define NOISE     100  // Noise/hum/interference in mic signal
#define SAMPLES    60  // Length of buffer for dynamic level adjustment
#define TOP    (N_PIXELS +1) // Allow dot to go slightly off scale

// Comment out the next line if you do not want brightness control
#define POT_PIN     3  // if defined, a potentiometer is on Pin #3 (A3)   ❹

byte
  peak     = 0,   // Used for falling dot
  dotCount = 0,   // Frame counter for delaying dot-falling speed
  volCount = 0;   // Frame counter for storing past volume data

int
  vol[SAMPLES],   // Collection of prior volume samples
  lvl      = 10,  // Current "dampened" audio level
  minLvlAvg = 0,  // For dynamic adjustment of graph low & high
  maxLvlAvg = 512;

Adafruit_NeoPixel strip =
  Adafruit_NeoPixel(N_PIXELS, LED_PIN, NEO_GRB + NEO_KHZ800);

void setup() {
  memset(vol, 0, sizeof(vol));  // Clear the sample array
  strip.begin();
}

void loop() {
  uint8_t  i;
  uint16_t minLvl, maxLvl;
  int      n, height;

  n  = analogRead(MIC_PIN);       // Raw reading from mic   ❺
  n  = abs(n - 512 - DC_OFFSET);            // Center on zero
  n  = (n <= NOISE) ? 0 : (n - NOISE);      // Remove noise/hum
  lvl = ((lvl * 7) + n) >> 3;     // Dampened reading (else looks twitchy)

  // Calculate bar height based on dynamic min/max levels (fixed point)
  height = TOP * (lvl - minLvlAvg) / (long)(maxLvlAvg - minLvlAvg);
  if(height < 0L)         height = 0;   // Clip output
  else if(height > TOP) height = TOP;
```

```
    if(height > peak)     peak  = height; // Keep 'peak' dot at top

// if POT_PIN is defined, we have a potentiometer on Pin #3 on Trinket
    uint8_t bright = 255;

#ifdef POT_PIN
bright = map(analogRead(POT_PIN),0,1023,0,255);    ❻
#endif
    strip.setBrightness(bright);

    for(i=0; i<N_PIXELS; i++) {    ❼
      if(i >= height)
        strip.setPixelColor(i, 0, 0, 0);
      else
        strip.setPixelColor(i,Wheel(map(i,0,strip.numPixels()-1,30,150)));
    }
    strip.show(); // Update strip

    vol[volCount] = n;                      // Save for dynamic leveling
    if(++volCount >= SAMPLES) volCount = 0; // Rollover sample counter

    // Get volume range of prior frames    ❽
    minLvl = maxLvl = vol[0];
    for(i=1; i<SAMPLES; i++) {
      if(vol[i] < minLvl)      minLvl = vol[i];
      else if(vol[i] > maxLvl) maxLvl = vol[i];
    }
    if((maxLvl - minLvl) < TOP) maxLvl = minLvl + TOP;
    minLvlAvg = (minLvlAvg * 63 + minLvl) >> 6; // Dampen min/max levels
    maxLvlAvg = (maxLvlAvg * 63 + maxLvl) >> 6; // (fake rolling average)
}

uint32_t Wheel(byte WheelPos) {    ❾
  if(WheelPos < 85) {
    return strip.Color(WheelPos * 3, 255 - WheelPos * 3, 0);
  } else if(WheelPos < 170) {
      WheelPos -= 85;
      return strip.Color(255 - WheelPos * 3, 0, WheelPos * 3);
  } else {
    WheelPos -= 170;
    return strip.Color(0, WheelPos * 3, 255 - WheelPos * 3);
  }
}
```

❶ Change this value if your project has a different number of NeoPixels.

❷ If you change the pin for the microphone, change this value. Use the analog pin number, not the Trinket (digital) pin number, per "Connectivity" on page 5.

❸ The NeoPixel pin can be changed to any digital pin.

❹ A potentiometer to adjust the pixels is recommended. The potentiometer center lug is connected to an analog pin. Because this uses analog 3 (Trinket pin #3), you must remove the Trinket to program it (pins #3 and #4 are shared with the USB port).

❺ The value from the microphone is read here. It is then conditioned such that low volumes tend to turn pixels off, while loud, continuous volumes of sound make most of the pixels light.

❻ If a potentiometer is connected, read it and use the value to set the brightness of the LEDs.

❼ Color pixels are based on a rainbow gradient. If you want other color schemes, you can change the pixel color setting algorithm here.

❽ minLvl and maxLvl indicate the volume range over prior frames, used for vertically scaling the output graph (so it looks interesting regardless of volume level). If they are too close together, though (e.g., at very low volume levels), the graph becomes coarse and "jumpy," so some minimum distance is kept between them (this also lets the graph go to zero when no sound is playing).

❾ An input value of 0 to 255 returns a color value that transitions from R to G to B and back to R.

Adjustments

The main adjustment you will want to make is to the gain on the microphone breakout, which is done using a tiny silver potentiometer on the back of the board. Use a small Phillips screwdriver to make small adjustments while you make sounds or play music with both some loud and soft passages. This might take a bit of trial and error.

You may want to rearrange your pixels to produce colors in a pattern that you like. You can also change some of the constants at the beginning of the program to adjust the behavior. Finally, the brightness potentiometer is optional. In a cabinet, you may want maximum brightness. In a dark room, it is beneficial to be able to tone down the light a bit.

Mounting

The typical color organ cabinet of the 1970s had a wood grain or black plastic box and a clear diffuser. Of course, the wood grain was typically faux.

To create your own cabinet, you can choose nearly anything. A clear plastic case works well, but the light will not diffuse through a clear lid; it will go straight through and you will not get that fuzzy-light look.

You may select a cabinet size to suit your decor. Repurposing a box made of nearly any material is ideal. To give it that faux-wood-grain look, there are a number of contact papers marketed as shelf liners that would do nicely. To make an inexpensive diffuser, a replacement fluorescent light fixture plastic cover is ideal and inexpensive.

You can find supplies at home stores. If you want a more professional look, a well-made wood cabinet (Figure 5-8) is hard to beat. Repurposing an old speaker cabinet or other box is ideal.

Figure 5-8. *Mounting the project inside a wooden box with diffuser*

Place your electronics in the back of the box, spacing your LEDs to suit your desired pattern. Test the project before securing the LEDs to the back of the box to ensure you like the light pattern when sound is made. If you find the sketch is not producing the ideal light pattern, you can change some of the parameters to get a reaction more suited to your taste. Ensure you adjust the microphone gain to pick up the sound at the levels you want. When done, cover the front with the diffuser, and place it in the desired area to add that special ambience.

Kaleidoscope Goggles

This is a very popular project from Adafruit by its resident blinky expert Phil Burgess. Wearable electronics are exploding in high fashion, trendy adornments, and cosplay. When it comes to costumes, a glowy set of steampunk goggles is irresistible. Note that these goggles are for wearing on top of your head (on a hat is perfect) as a fashion accessory—they will probably be too hard to see out of with all the electronics added.

Adafruit 16-pixel NeoPixel LED rings fit perfectly inside the eyecups of most 50 mm round goggles (which is a very common size). It is almost as if these rings were made with this project in mind (actually, they were, then clever folks came up with many more uses). This project is a bit more difficult than the previous ones, but can be done with some patience.

Figure 5-9. *Kaleidoscope goggles by Phil Burgess*

Parts List

- Trinket 3V, Adafruit #1500 or Maker Shed #MKAD70 (an Adafruit Gemma #1222 would work as well and use the same code as the Trinket)
- 16-NeoPixel ring, Adafruit #1463 or Maker Shed #MKAD75
- 3.7V 150 mAh lithium polymer battery, Adafruit #1317, and USB LiPo battery charger, Adafruit #1304
- Alternate power source: 3xAA battery holder, Adafruit #771 or similar, and 3 AA batteries
- Heat-shrink tubing pack, Adafruit #344
- Costume goggles (50 mm / 2 inch round), Adafruit #1577 or similar
- Connecting wire (20 to 26 gauge, stranded or solid)
- JST-PH battery extension cable (500 mm), Adafruit #1131

  OR

  Optional: JST-PH surface mount connector, Adafruit #1769 (for Trinket version 1.1)

Tools

You will need the following tools to build this project:

- A soldering iron to make the wire connections
- Adhesive, such as tape for a temporary fit or hot glue for final assembly
- Wire strippers to cut wires and remove insulation
- Pliers to help make wire splices

- Optional: Heat gun for heat shrink (be careful not to melt other parts)

Battery Selection

Choose one of the following to fit your budget and desired run time. The pros and cons of each are listed:

Lithium polymer (LiPo)
- A 150 mAh LiPo battery is tiny and easily fits within the goggles, but the power capacity is limited and run time will be shorter. To improve run times, the software can be changed to lower the overall LED brightness and reduce power usage, or you can substitute a larger LiPo battery. If the battery is too big, however, it will not fit inside the goggles and you will need to run wires and hide the battery elsewhere.
- Costs more initially, but it is rechargeable.
- You will also need a LiPo charger and JST socket inside the goggles. For the latter, cut a LiPo battery extension cable in half or use the JST surface mount connector. The battery gets disconnected for charging (it does not charge in place).

3 AA alkaline cells
- This saves money initially; the battery case and cells are inexpensive and you do not need a charger.
- Provides excellent run time. You can use brighter, showier LED patterns, and you can easily swap in a fresh set of batteries.
- The battery pack is much larger and heavier. It will not fit inside the goggles—you will need to run wires and put the battery pack in a pocket or conceal it behind (or within) a hat, mask, pants, or shirt.

Wiring

You need only a few connections, with a couple of tricky things to watch out for.

Depending on the eyewear design, you might need to snake wires through some parts, and you may not have an opportunity to test the full circuit separately on the bench first. If you do build the complete circuit first and install in the goggles afterward, make sure all the wires are long enough (e.g., to go across the bridge of the nose); extra wire can always be folded up, but wires that are too short are frustrating, as they would need to be replaced.

As seen in Figure 5-10, the wires pass through holes in the sides of the goggle eyepieces. Therefore, it is not easy to build the circuit, test it, and then mount it in the goggles. It must be built around the goggles and later folded into place. This can be tricky!

You may need to split power leads three ways from the battery to the Trinket and the two NeoPixel rings. Even a regular 1-to-1 inline splice can be tricky for the inexperienced; three-way is an extra challenge. It is a little easier with narrow-gauge wire (e.g., 26 gauge). Referring to Figure 5-11, work slowly and methodically. Remember to slide heat-shrink tubing on **first**, before joining the wires, and use proper soldering techniques. Heat the wires and apply the solder there; do not move a glop of solder from the iron onto the connection.

The plus and minus (power and ground, respectively) split three ways, from the battery holder (or JST socket) to the Trinket and NeoPixel ring(s), or two ways, if you're using a single monocle ring. Plus (+) connects to BAT + on the Trinket and positive power on the ring(s). Minus (-) connects to GND on both the Trinket and the rings.

Connect Trinket pin #0 to the Din connection on the first ring.

If using two rings, connect Dout from the first ring to Din on the second ring, as seen in Figure 5-12.

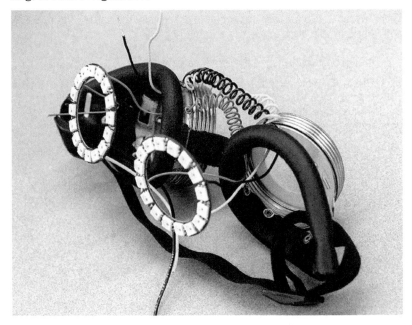

Figure 5-10. *Passing wires through the goggles*

Figure 5-11. *The dreaded three-way inline splice with heat shrink*

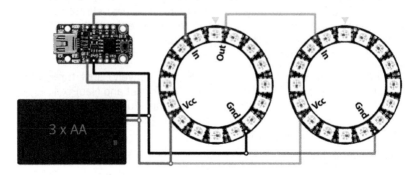

Figure 5-12. *The goggle wire layout using an AA battery pack*

When using a LiPo battery the circuit is essentially the same (Figure 5-13), with just a couple of changes:

- A LiPo battery replaces the 3xAA battery holder.
- Use a JST socket (cut from a battery extension cable) or the Trinket rear surface mount JST connector rather than soldering directly to the battery leads. This allows the battery to be unplugged for recharging.

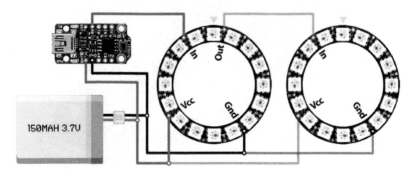

Figure 5-13. *The goggle wire layout using a LiPo battery*

Your LED code might not require any specific orientation to the rings, in which case the rings can be installed any which way. Other code requires a known direction for certain visual effects to work (e.g., eye blinks). If aiming for the latter, try to follow the orientation shown here, and make sure the wire lengths are sufficient for those angles. Viewed from the front, the first NeoPixel (#0) should be at the top—immediately to the *left* of the Dout connection. Viewed from the back, it will be on the *right* of that hole. You can make a mark on the back with a marker to help line things up when installing the circuit into the goggles.

Software

Be sure you install the Adafruit_NeoPixel library introduced in Chapter 4.

The standard goggles code can be found at *File→Sketchbook→Libraries→Adafruit_NeoPixel→goggles* and in the repository for this book (*http://bit.ly/GettingStartedWithTrinket*) (directory *Chapter 5 Code*, subdirectory *Chapter5_02Goggles*).

From the IDE Tools→Board menu, select Adafruit Trinket 8 MHz. Connect the Trinket to your computer with the USB cable, press the reset button on the Trinket board, and then click the upload button in the Arduino IDE (a right arrow icon). If all is well, when the battery is connected, you should get a light show from the LEDs.

While it is very showy, the goggles example sketch included with the library requires a great deal of power. It will drain the small LiPo battery in 15 minutes or less. This is why, alternatively, you can use the larger external pack with code.

Or, to save power, you can write your own program that lights fewer LEDs at a time. Sticking with primary colors (red, green, or blue) also can reduce power use. (For instance, white requires about three times as much current, because all three colors are lit simultaneously.) A simpler version, shown in Example 5-3, is still interesting while using just a small fraction of the power of the goggles sketch included with the library (it can run for about three hours on the small LiPo battery).

Example 5-3. NeoPixel Goggles code

```
/* Low-power NeoPixel goggles example.
   Makes a nice blinky display with just a few LEDs on at any time.
*/
#include <Adafruit_NeoPixel.h>     ❶

#define PIN 0          // NeoPixel rings signal connect to Trinket Pin #0
#define NUMPIXELS 32 // 16 LEDs per ring, 32 for two rings

Adafruit_NeoPixel pixels = Adafruit_NeoPixel(NUMPIXELS, PIN);
```

```
uint8_t  mode    = 0, // Current animation effect    ❷
         offset  = 0; // Position of spinny eyes
uint32_t color   = 0xFF0000; // Start red
uint32_t prevTime;

void setup() {         ❸
  pixels.begin();
  pixels.setBrightness(85); // 1/3 brightness (range: 0-255)
  prevTime = millis();
}

void loop() {          ❹
  uint8_t  i;
  uint32_t t;

  switch(mode) {       ❺

    case 0: // Random sparks - just one LED on at a time!
      i = random(NUMPIXELS);
      pixels.setPixelColor(i, color);
      pixels.show();
      delay(10);
      pixels.setPixelColor(i, 0);
      break;
    case 1: // Spinny wheels (8 LEDs on at a time)
      for(i=0; i<16; i++) {
        uint32_t c = 0;
        if(((offset + i) & 7) < 2) c = color; // 4 pixels on...
        pixels.setPixelColor(   i, c); // First eye
        pixels.setPixelColor(31-i, c); // Second eye (flipped)
      }
      pixels.show();
      offset++;
      delay(50);
      break;
  }
  t = millis();
  if((t - prevTime) > 8000) { // Every 8 seconds...    ❻
    mode++;                    // Next mode
    if(mode > 1)\{             // End of modes?
      mode = 0;                // Start modes over
      color >>= 8;             // Next color R->G->B
      if(!color) color = 0xFF0000; // Reset to red
    }
  for(i=0; i<NUMPIXELS; i++) pixels.setPixelColor(i, 0);    ❼
    prevTime = t;
  }
}
```

❶   You must include the Adafruit-NeoPixel library, which you can get
     from *https://github.com/adafruit/Adafruit_NeoPixel* and install per
     the instructions in Chapter 4.

❷   Initialize the variables mode and color for lighting effects.

❸ The **setup** function initializes the data for the NeoPixel rings in **strip**. The brightness is set here also, and you can vary the brightness from off (0) to full (255). You'll save battery life if you keep the brightness down (and NeoPixel LEDS are very bright at 255).

❹ For different animations, change the **loop** function code.

❺ Two modes are preprogrammed: sparks (single pixels) and spinny wheels (eight LEDs at one time).

❻ Currently, the animation mode is changed every 8 seconds. On every mode change, the color is shifted through red, green, and blue.

❼ This final loop sets the colors on the rings before looping back.

Final Assembly and Use

NeoPixel LEDs are very bright and focused. You will probably want to create some form of diffuser to soften the light. If the goggles you chose were originally designed for welding, they will have very dark ultraviolet (UV) filters installed, usually with a second clear glass or plastic lens over this. Unscrew each eyepiece and remove the welding filters. Leave the goggles disassembled.

For a simple diffuser, set the lens on a piece of paper and outline it with a pencil or pen. Regular copier or printer paper works fine, or you can use fancy drafting vellum if you have it (Figure 5-14). Cut out your traced shapes with scissors. Vellum is very translucent, so you might want two layers per eye (four circles total).

Figure 5-14. *Optional vellum diffuser inserts for the goggles*

If you have access to a laser cutter (through a local hackerspace, your school, or elsewhere), you may measure the diameter of the lenses removed from your goggles as a template for cutting new ones; 1/16" white acrylic works well for this.

Reassemble the goggles using just the clear lenses with the diffusers behind them. Connect the battery temporarily to make sure all the electronics are working prior to final assembly. It is easier to troubleshoot while everything is out in the open. If your goggles are made of metal, make sure there is no contact between the goggles and any exposed conductors.

Fit the NeoPixel rings in place inside each eye cup. As explained in the wiring section earlier, there is a definite "up" orientation to the rings. Make sure they go in the right way. The rings can then be held in place using a few dabs of hot glue around the perimeter.

Disassembly

If you need to remove the rings later, dip a Q-tip in rubbing alcohol, touch it to the edge of each blob of hot glue and allow it to soak in for a few seconds. This does not dissolve the glue. It seeps between the two parts and cleanly breaks the bond. The glue should peel away with little effort.

Ring removal after having been positioned with a glue gun:

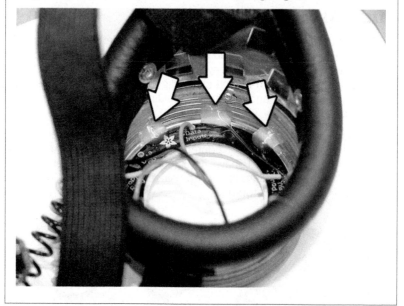

Next, position the Trinket board inside one of the eye cups and secure it with hot glue as well (Figure 5-15). You may want to position it where the USB port is accessible, so you can upload new code later. Then fold up any wire slack and hold it in place with a few more dabs of glue.

Because the goggles used in this example are metal, rest the Trinket board on a piece of foam tape to avoid shorting. Plastic goggles are easier to work with in this regard.

If you're using a small LiPo battery, you will probably want to use masking tape (not hot glue) to hold the battery in place in one of the eye cups. Then it can easily be removed for recharging. If you are using an external battery pack, sew a few loops of thread to secure the wire to the goggle strap as a strain relief.

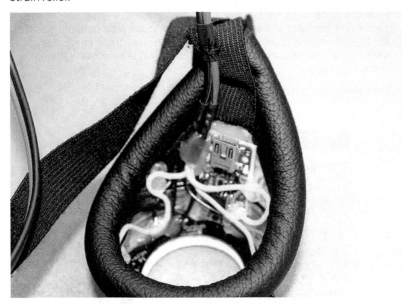

Figure 5-15. *Positioning the Trinket inside the goggles*

Safety and Common Sense

Your LED goggles are a fashion accessory. They should be worn on your forehead or on a hat, not over your eyes. The scattered light inside the goggles is still very bright, and can cause headaches or possibly eye injury or nausea. The goggles may also limit your peripheral vision, so wearing them is not a great idea: stick them up above your eyes.

If you have modified a pair of welding or safety goggles, their design is now compromised, and they should no longer be used for welding or safety. For the same reasons, if attending an event like Burning Man, take one (or several) pairs of "real" sealed dust goggles in addition to your LED "fun" goggles. Do not rely on decorative goggles for protection.

Safety Checklist
1. Never use the altered goggles for safety or eyewear. They are strictly a fashion item.
2. If your goggles have metal frames, make sure there is no contact with exposed conductors on the Trinket board, NeoPixel rings, or wires.

Wearable Electronics

Wearable electronics has become the fastest growing segment of the consumer and hobbyist industry in the last two years with no end in sight. Smart watches and fitness bands dominate the current commercial space, while talk of Google Glass and New York Fashion Week electronic creations crowd the media.

The Pebble Watch (left) and Trinket Necklace (right):

Likewise, hobbyists around the world are creating some mind-blowing wearable creations. Seeing the potential, many companies have developed materials to accelerate the market. Adafruit, SparkFun, and other companies have established wearable groups or market wearable electronic components.

The Trinket provides for electronic programmability in a small package highly suitable for wearable projects such as watches, light-up clothing, and electronic jewelry.

Servos

Servo motors are a popular and simple way to add movement to projects. Best known to some for RC model control, servos, shown in Figure 5-16, are also used in many microcontroller projects.

Figure 5-16. *A standard servo with a variety of horns (mechanical connectors)*

A standard servo rotates 180 degrees (halfway around a circle). The standard model will not fully rotate like an electric motor. Another type of servo, a full-rotation servo, is designed to rotate a full 360 degrees.

Servos are most often packaged with *horns*, various shaped pieces that you can screw on top of the servo shaft. This makes it easier to mechanically connect the servo to the item you want to move. You can drill the various radial holes on a horn to fit screws or other connectors.

Inside a Servo

A servo actually has a rather complex series of components inside. Besides a motor, it has a digital circuit that detects specific pulses on the data line to determine the angle to turn to. An internal potentiometer (variable resistor) gives the circuit feedback on where it is in the rotation sweep. Gearing provides *torque* (mechanical power) to the rotation.

Servos typically have three connections: the digital data connection, power, and ground. Servos come in various voltages, but 5 volts is very common.

The servo expects to receive a pulse every 20 milliseconds (fast to us, not so much to an 8 MHz microcontroller), as shown in Figure 5-17. If the pulse width (the time it is high/on) is 1 millisecond, then the servo will go to the 0 degree position. As the pulse width increases, the shaft will turn. A width of 1.5 milliseconds produces a 90-degree rotation, and a 2-millisecond (maximum) value produces the full 180-degree rotation.

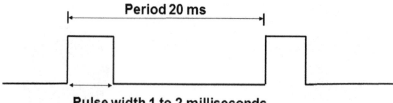

Pulse width 1 to 2 milliseconds

Figure 5-17. *Pulse width stream used to control a servo*

Trinket Servo Control

The Trinket's small size makes it ideal for lightweight projects, including robotics. The project shown in Figure 5-18 demonstrates the use of a standard hobby servo with the Trinket.

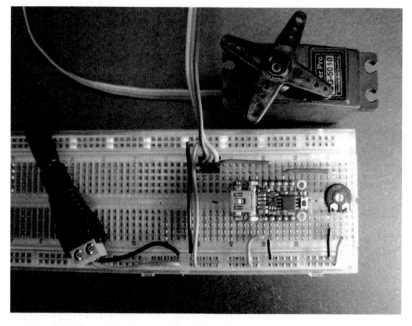

Figure 5-18. *Controlling a servo with Trinket*

The standard Arduino IDE Servo library will not work with 8-bit AVR microcontrollers like the ATtiny85 on the Trinket due to differences in the chip's timer hardware.

The Adafruit_SoftServo library, which we met in Chapter 4, uses one of the two timers available on the Trinket to provide servo control pulses.

To start using servos, you'll build a simple circuit that allows you to control the servo with a potentiometer.

Parts List

- Trinket 5V, Adafruit #1501 or Maker Shed #MKAD69
- USB cable for power and reprogramming
- Standard 5V hobby servo, Adafruit #155, #169, or similar (such as Maker Shed #MKPX17)
- Potentiometer, anything from 1K ohms to 10K ohms, Adafruit #356, #562, or similar
- Half breadboard, Adafruit #64, Maker Shed #MKKN2, or similar
- Breadboard hookup wires, Adafruit #153, Maker Shed #MKSEEED3, or similar
- 5V power supply, Adafruit #276 or similar
- Female 2.1/5.5 mm DC power connector, Adafruit #368

Wiring

The connections are shown in Figure 5-19. The Trinket is connected to the power rail, as is the servo and the outer pins of the potentiometer. The signal wire of the servo goes to Trinket pin #0, while the center of the potentiometer is wired to Trinket pin #2 (analog pin 1).

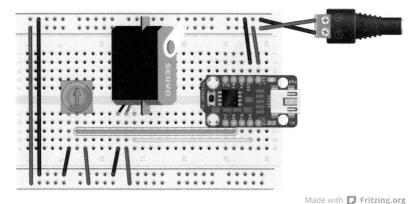

Made with **Fritzing.org**

Figure 5-19. *Wiring diagram for Trinket control of a servo*

Code

The project code is in Example 5-4 and can be downloaded from the repository for this book (*http://bit.ly/GettingStartedWithTrinket*) (directory *Chapter 5 Code*, subdirectory *Chapter5_05Servo*).

Example 5-4. Code for the Trinket Servo Control project

```
/* Trinket Servo Control Sketch */

#include <Adafruit_SoftServo.h>
#define SERVO1PIN 0     ❶
#define POTPIN 1        ❷

Adafruit_SoftServo myServo1; // create servo object

void setup() {
  OCR0A = 0xAF;                   ❸
  TIMSK |= _BV(OCIE0A);

  myServo1.attach(SERVO1PIN);     ❹
  myServo1.write(90);             ❺
  delay(15);                      // wait 15 ms for servo to reach position
}

void loop() {
  int potValue; // variable to read potentiometer
  int servoPos; // variable to convert voltage on pot to servo position
  potValue=analogRead(POTPIN);    ❻
  servoPos = map(potValue, 0, 1023, 0, 179);
  myServo1.write(servoPos);       ❼
  delay(15);      // wait 15 ms for servo to reach position
}

volatile uint8_t counter = 0;  // timer counter variable

SIGNAL(TIMER0_COMPA_vect) {       ❽
  // this gets called every 2 milliseconds
  counter += 2;
  // every 20 milliseconds, refresh the servo!
  if (counter >= 20) {
    counter = 0;
    myServo1.refresh();
  }
}
```

❶   The servo control wire (usually orange) is connected to the Trinket on this pin number (pin #0).

❷   The potentiometer center pin is connected to this pin (Trinket pin #2, which is analog 1 on the ATtiny85).

❸ These two lines set up the interrupt that will refresh the servo every 20 milliseconds to keep it in the position desired.

❹ This calls the Servo library to identify that the servo is on `SERVO1PIN` on the Trinket.

❺ Here, we set an initial position for the servo (the servo could be pointing anywhere when you first power it up).

❻ Here, we read the voltage on the potentiometer (`analogRead` returns a value of 0 to 1023) and then map the value to a value of 0 to 179 degrees.

❼ This tells the servo to go to the desired position.

❽ The `SIGNAL(TIMERO_COMPA_vect)` function is the interrupt that is called by the microcontroller every 2 milliseconds. The built-in `millis` timer function keeps track of time, and when 20 milliseconds has elapsed, this function will refresh the servo.

The code uses 1,678 bytes of the 5,310 maximum. This leaves a good amount of room for user code (lights, robotics; the sky is the limit).

Use

As the potentiometer is turned, the servo will rotate from 0 to 180 degrees. The potentiometer is used as a voltage divider, changing the voltage on the middle pin from 0 to 5 volts through the sweep of the knob. This is read on analog pin 1 (which is Trinket pin #2). The `analogRead` value is mapped (linearly interpolated) to a number from 0 to 179, and we send that to the Servo library.

You can expand this circuit for a number of useful projects. You'll use the Servo library again in Chapter 6.

Going Further

If you need to control more than two servos, you might consider a multi-channel Pulse Width Modulation servo controller. Adafruit sells an I2C-controlled model (#815) that controls 16 servos (chainable to a staggering 192 servos). To date, the library is not tested for Trinket. A larger microcontroller may be needed for some servo applications.

Analog feedback servos have an extra wire coming from the servo body. This is connected to the position-sensing potentiometer to provide analog information on where the servo is for your own use. A typical application is having a user move an object while the microcontroller records the analog values; then the program can *play back* those same movements. See *http://learn.adafruit.com/analog-feedback-servos* for more information on these specialized servos.

Using I²C—The Two-Wire Interface

Besides serial communication, Trinket is able to use the USI interface for performing *Inter-Integrated Circuit* (I²C) connectivity. I²C is also known by the generic name *Two-Wire Interface* (TWI). I²C uses only two bidirectional signal lines, Serial Data Line (SDA) and Serial Clock Line (SCL), pulled up with resistors. You can communicate using typical voltages for Trinket (e.g., 5 volts for Trinket 5V, 3.3 volts for Trinket 3V).

A communication channel is established between the controlling unit (most often a microcontroller like Trinket), which is designated the *master*, and a controlled object called the *slave*. The master is responsible for generating the Clock signal. Communication may be bidirectional, as both master and slave can send and receive data.

You can connect multiple devices, as shown in Figure 5-20. Each slave device has a fixed address. The master uses these addresses to select the slave with which it wishes to establish communication. Older devices have a 7-bit address, while newer devices have a 10-bit address. Even with the older devices, 7 bits supports 128 devices, which is probably more than we would ever load a Trinket with. The device manufacturers often either *hard-code* an address in a device or allow you to select the address (from a limited range) by changing jumpers. If you wish to have many devices of the same type (displays, sensors), you might be limited by the address selections the device manufacturers have provided.

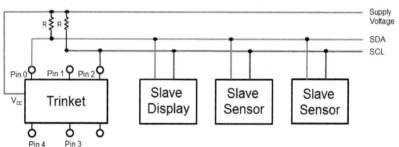

Figure 5-20. *Typical I²C connections—you can connect different slaves to the bus*

The pull-up resistors (R) should have values between 2.2K and 10K ohms. Depending on the circuit, the resistors may not be needed (some breakout boards place the resistors in the circuit for the user). But if communication issues arise, the lack of resistors is the most likely cause. Connect each I²C line to a resistor, with the other end connected to the positive voltage line.

A number of useful devices, such as displays, memory, sensors, and more, come with I²C interfaces. I²C provides a robust communication method and only uses two digital pins. One limitation on the Trinket is that the two pins must be pins #0 and #2 (as the software uses the ATtiny85 USI to provide the communication hardware).

I²C Software

Larger Arduino compatibles communicate over the I²C bus using the *Wire* library. This provides an intermediate layer of code so you don't need to know the low-level details of sending and receiving I²C commands over the hardware interface. Since the Trinet does not have dedicated I²C hardware (using the USI instead), the standard Wire library doesn't work. Fortunately, there is an alternative Wire library for the ATtiny85: *TinyWireM* (the *M* stands for master; there is also a TinyWireS library for slave communication). TinyWireM provides similar functionality for most projects requiring I²C connectivity via Wire function definitions.

Some code, either in third-party libraries or in sample programs, may require changes to make calls to TinyWireM instead of Wire. Some libraries make this switch in code transparently; other libraries are rewritten to use TinyWireM. In a library file or program, you can make the change by defining the following lines:

```
#include <TinyWireM>
#define Wire TinyWireM
```

To edit a library file, it is easiest to use a text editing program navigate to your library folder and open the library files, which often have file extensions of *.c* (C source code), *.cpp* (C++ source code, often mostly consisting of C code), and *.h* (header files, defining variables and drawing in other code via #include statements like the one you just saw). If you were to manually change all Wire calls to TinyWireM, you would need to edit the C code files. If you use the #define Wire TinyWireM trick in a header file, you may not need to edit the rest of the source. A great deal depends on how the library authors implemented their code.

If you get errors similar to "'Wire' is not defined", then you have not made all of the changes to use TinyWireM instead of Wire.

Using I²C Displays

The I²C bus on the Trinet allows the connection of text displays such as those in Figure 5-21. You may grow tired of using a single blinking red LED to provide status feedback for a program, or cringe at using a serial connection back to your computer. Displays provide a logical way for standalone projects to provide user information (and catch people's eyes).

Figure 5-21. *Many different smart displays have I²C interfaces*

Displays provide information from a projects' hardware to the user. The next project provides information from a temperature and humidity sensor to the user via a display.

Displays may be subdivided into four groups: seven-segment LED numeric displays, LED matrix displays, character displays (often 16 characters by 2 lines, but there are others), and graphical displays that use pixels to display text and images. Most of these displays require many digital pins to control their functions. Chip manufacturers have developed I²C driver chips that allow for controlling the displays over the I²C bus, leaving the hard work to the interface chip. Electronics companies such as Adafruit design and build circuit boards, called *backpacks*, which contain the driver chip and I²C connections, plus the connections to the display. This adds some cost, but allows a small controller like Trinket to interface in the same way as a larger microcontroller.

The main limitation on displays is *buffering*. Many displays require the memory of the display contents to be handled by the microcontroller. This is especially true of graphical displays. With only 512 bytes of RAM on the Trinket, buffering graphical displays can be very difficult. If you are contemplating a design with a large, unbuffered LCD display and a Trinket, consider using another microcontroller. But for many types of displays, the Trinket provides the capability at the right price point.

Using an I²C display requires the TinyWireM library plus a library to provide the low-level communication between the Wire protocol and the display backpacks. As the driver chips on the backpacks may vary, check which display-dependent library you need for the display used in your project.

Temperature and Humidity Sensing

Internet of Things (IoT) projects often read sensors and report their data to a monitoring display. Temperature and humidity sensors are common in IoT sensing. The DHT series of sensors allow for measuring both temperature and humidity in one package.

The next project (seen in Figure 5-22) provides environment sensing with a DHT sensor. It may be built and placed in a very small enclosure, allowing for environmental monitoring in many different scenarios. You can use the code and concepts in a number of other projects.

Figure 5-22. *The Temperature and Humidity project on the breadboard*

Parts List

- Trinket 5V, Adafruit #1501 or Maker Shed #MKAD69 (do not substitute a Trinket 3V on this project)
- I²C character LCD backpack, Adafruit #292
- Monochrome or color LCD display compatible with the LCD backpack, such as the Adafruit #181 monochrome or #398 or #399 color displays
- DHT22 temperature and humidity sensor, Adafruit #385 (you can also consider the DHT11, Adafruit #386, with a change in the code).
- Half breadboard, Adafruit #64, Maker Shed #MKKN2, or similar (full breadboard just fine)
- Breadboard wires, Adafruit #153, Maker Shed #MKSEEED3, or similar
- 5V power supply, Adafruit #276 or similar
- Female 2.1/5.5 mm DC power connector, Adafruit #368

- 1,000-ohm resistor—the DHT22 comes with a 10K-ohm pull-up resistor, but this is too weak for use on Trinket pin #1, so you will not use it unless you decide to change the circuit

Libraries

We'll use the following libraries to communicate with these components (see "ATtiny-Optimized Libraries" on page 39 for library locations and "Installing Libraries" on page 41 for instructions on installing libraries):

- TinyWireM library (for I^{2}C communication)
- TinyLiquidCrystal library
- TinyDHT library

The TinyLiquidCrystal library is the Adafruit LiquidCrystal library with additional support for one of the display driver chips, along with use of TinyWireM instead of the Wire library. The TinyDHT library is special. The Adafruit DHT library uses floating-point (decimal) math, which loads more code than the Trinket can handle. To save space, TinyDHT uses integer math, resulting in return values that are rounded to the nearest integer. Note the Arduino LiquidCrystal library is not used for Trinket. For Adafruit displays, it is best to check their thorough tutorials (*https://learn.adafruit.com/*) for library requirements.

The LCD Display

Adafruit carries many character LCD display varieties with multiple sizes and backlight colors. The backpack used in this build has 16 characters per line, with two lines.

The Adafruit I^{2}C/SPI character LCD backpack, as shown in Figure 5-23, lets you control these displays by sending data over the two-wire I^{2}C interface. Standard LCDs require a large number of digital pins, straining the capability of even an Arduino Uno. Use of the I^{2}C backpack considerably reduces the number of pins needed.

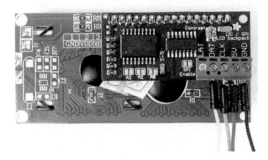

Figure 5-23. *The Adafruit I²C backpack soldered to a 16x2 display*

You must assemble the I²C backpack according to the instructions on the Adafruit website (*http://learn.adafruit.com/i2c-spi-lcd-backpack/assem bly*). The backpack is then placed on the back of the display.

For color displays, there are three backlight connection pins: display pins 16, 17, and 18 control the three color backlights. If you connect pin 16 from the backpack to pin 16 on the display, the I²C controls will adjust the red light. You can place a jumper from one or more of the backlight pins to the backpack's pin 16, and the software will vary the colors you've connected with a jumper. You should make your color choice before soldering on the backpack. Alternatively, rather than a color display, you can choose a "classic" blue and white 16x2 LCD. The monochrome display only has one backlight control pin, making it more straightforward to control.

Once you know which pins are needed for the backlight, solder the backpack to the display. The completed display is connected to the project per the wiring diagram in Figure 5-24. To be sure the project has enough power, use an external 5-volt supply. It must supply at least an ampere (1,000 mA). Wire the DAT pin on the backpack to Trinket pin #0 and the CLK pin to Trinket pin #2. Connect the backpack 5V to the project common 5V power bus and GND to ground.

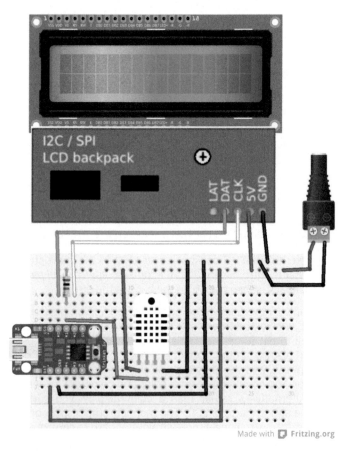

Made with **F** Fritzing.org

Figure 5-24. *The Temperature and Humidity wiring diagram*

Testing the Display

To test the LCD, use the Hello World sketch in Example 5-5 (also available in the repository for this book (*http://bit.ly/GettingStartedWithTrinket*), directory *Chapter 5 Code*, subdirectory *Chapter5_06DisplayTest*). It is important to test the display *before* adding the sensor to isolate any errors up to this point in the project.

Example 5-5. I²C display test program

```
/* Test sketch for Adafruit I2C/SPI LCD backpack and a character display */

#include <TinyWireM.h>            ❶
#include <TinyLiquidCrystal.h>
```

```
TinyLiquidCrystal lcd(0);   ❷

void setup() {
  lcd.begin(16, 2);      ❸
  lcd.setBacklight(HIGH);
  lcd.print("hello, world!");
}

void loop() {         ❹
  lcd.setCursor(0, 1);
  lcd.print(millis()/1000); // print the number of seconds since reset
}
```

❶ Include the Wire library for the ATtiny85 and the display libraries.

❷ Call the display library, default I²C address #0 (backpack pads A0–A2 not jumpered). If the address is different, change the number to the correct address.

❸ The display is initialized as a 2-line, 16-character display (change to suit your display as needed), then the backlight is turned on, and the message is written to the first line.

❹ Set the cursor to column 0, line 1 (note: line 1 is the second row, since counting begins with 0).

Before running the test program, be sure both the TinyWireM and Adafruit TinyLiquidCrystal libraries are installed (as described in "Installing Libraries" on page 41. If there are errors indicating that the library or function code cannot be found, be sure to review Chapter 4 on where to locate libraries and be sure the names of the directories match up with the library names.

Adjustment

When the code is running, text should be on the display. If there is no text, your first adjustment should be to the small silver potentiometer on the rear of the backpack. Adjust this potentiometer with a small screwdriver until the intensity of the text is readable.

If the text is still not displaying and the backlight is off, check the wiring from the backpack to the breadboard, check power (is the Trinket's power LED on?), then check the soldering on the backpack to ensure all connections are correct.

Sensing

Once the display is working with the Trinket, you can expand the project. Next, we'll add a temperature sensor (Figure 5-25).

Figure 5-25. *The DHT11 and DHT22 sensors*

Connect the DHT22 to power and ground, as shown in Figure 5-24. The DHT22's digital data line connects to Trinket pin #1. The sensor is designed such that the data pin must be pulled high via an external resistor.

Using Trinket pin #1 for a signal input requires a different pull-up resistor value than the 10K-ohm resistor that comes with the sensor from Adafruit. Because of the Trinket's onboard 470-ohm resistor/LED on pin #1, you need a lower resistance value to "pull more strongly" (provide a more balanced voltage divider). If there are signal strength issues with the digital pin, make sure the sensor data line has adequate pull-up. Pin 3 on the sensor may be ignored; it is not used.

Remember to program your Trinket out of circuit because its communication pins are shared with data pins.

The DHT22 provides a 0–100% humidity reading with 2–5% accuracy and −40 to 80°C temperature readings with ffl0.5°C accuracy. More detailed specifications and use documentation may be found at *http://learn.adafruit.com/dht*.

To get the timing correct on the DHT sensor, the Trinket is clocked to 16 MHz (the same as the Arduino Uno) in the first line in the **setup** routine. This requires the 5V Trinket, as the 3V Trinket is not guaranteed to function at 16 MHz.

Change your board type in the Arduino IDE to Trinket 5V 16 MHz before compiling (via the Tools→Board menu item).

Code

The code for this project is in Example 5-6 and can be downloaded from the repository for this book (*http://bit.ly/GettingStartedWithTrinket*), (directory *Chapter 5 Code*, subdirectory *Chapter5_07DHT*).

Example 5-6. Code for the Temperature and Humidity Sensing project with I2C display

```
/* Trinket I2C Display and DHT Sensor Sketch */

#include <TinyWireM.h>        // Wire/I2C library for Trinket    ❶
#include <TinyLiquidCrystal.h> // LiquidCrystal using TinyWireM
#include <TinyDHT.h>           // Lightweight DHT sensor library
#include <avr/power.h>   // Needed to up clock to 16 MHz on 5v Trinket

#define DHTTYPE DHT22   // DHT 22 (AM2302)    ❷
#define TEMPTYPE 1       ❸
#define DHTPIN 1         // The sensor is connected to pin #1

DHT dht(DHTPIN, DHTTYPE);   ❹
TinyLiquidCrystal lcd(0);   ❺

void setup() {
  if (F_CPU == 16000000) clock_prescale_set(clock_div_1);   ❻
  dht.begin();              // Start temperature sensor
  lcd.begin(16, 2);         ❼
  lcd.setBacklight(HIGH);
}

void loop() {
  int8_t h = dht.readHumidity();               ❽
  int16_t t = dht.readTemperature(TEMPTYPE);

  lcd.setCursor(0, 0);
  if ( t == BAD_TEMP || h == BAD_HUM ) {   ❾
    lcd.print("Bad read on DHT");          ❿
  } else {
    lcd.print("Humidity: "); // Write values to LCD
    lcd.setCursor(10, 0);  lcd.print(h);
    lcd.setCursor(12, 0);  lcd.print(" % ");
    lcd.setCursor( 0, 1);  lcd.print("Temp:");
    lcd.setCursor( 7, 1);  lcd.print(t);
    lcd.setCursor(10, 1);  lcd.print("*F");
  }
  delay(2000); // Read values every 2 seconds (2000 milliseconds)   ⓫
}
```

❶   Load the libraries needed for the program.

❷   The sensor type can be defined as DHT11, DHT21, or DHT22. Set it for the one you are using.

❸   Define the variable for temperature scale: 1 for Fahrenheit, 0 for Celsius.

❹   Initialize the temperature sensor data structure.

❺   Initialize the display connected via I²C, default address 0 (backpack pins A0–A2 *not* jumpered)

❻   This special line sets the Trinket to run at 16 MHz.

❼   Define the display as 2 rows, 16 columns, and turn on the backlight.

❽   Read the humidity, then the temperature (note that the data sizes are different).

❾   The file *TinyDHT.h* defines the values for the variables that tell you when the sensor is reading funky values.

❿   If an error value is returned, display it on the LCD; otherwise, print the humidity and temperature on the display.

⓫   The sensor does not like reads faster than once every two seconds, per the datasheet (*https://learn.adafruit.com/dht*), so this line puts a delay in between each read.

The sketch compiles to 4,880 bytes of the 5,310 available. This leaves only 440 bytes of code for any additional functionality. Note that adding more display text will use the available space quickly. If you need decimal (floating-point) numbers, it will most likely exceed the code space available. The Arduino library functions to do floating-point math add up to 2,000 bytes of code. This is why the DHT library was forked to create the TinyDHT library—it uses integer math, which limits precision to one degree and one percent, but saves space.

How It Works

A couple of new concepts are introduced in this sketch. This is the first time in the book you've clocked a Trinket 5V up to 16 MHz (see the upcoming sidebar, "Running Trinket at 16 MHz" on page 96). The setup routine clocks the Trinket 5V up to 16 MHz, then initializes the sensor and display. The loop function reads the temperature and humidity and, if the values are valid, the sketch reports them on the LCD.

The red LED will probably glow softly as the digital sensor data flows into pin #1. This should not cause any problems.

You can heat up the sensor or cool it off to observe resulting temperature changes. For permanent use, select a weatherproof, sturdy enclosure and do not expose the display to moisture. Consider a box such as Adafruit #903 or #341 and keep the display indoors.

Troubleshooting

- If you get no display, go to the Hello World I²C sketch (in "Testing the Display" on page 90) and ensure that the display works.

- If you have no display running Hello World on the I²C backpack, use the contrast knob to change the LCD display contrast to a readable level. If you decided to use an external potentiometer to change contrast per the assembly instructions and not pin 16 on the backpack, use that.

- If you get a `Bad Read on DHT` error, the sensor is not talking to the Trinket correctly. Check the wiring and ensure there is a 1,000-ohm resistor from Trinket pin #1 to 5V. If you have an oscilloscope, you can look at the signal on pin #1 to ensure good high to low transitions given the 1K pull-up and the onboard LED. Try using a 10K-ohm pull-up resistor on pin #4, or wire the DHT signal line to Pin 4 and change the value of `DHTPIN` to 4. If you use pin #3 or #4, disconnect the wiring on the pin before you upload code, after which you can reconnect the wires.

- If you get 0% Humidity and 32 degrees F/ zero C, ensure Trinket 5V 16 MHz is selected as the Board type in the Arduino IDE Tools menu. The sensor code will not give correct readings on a Trinket 3V or at 8 MHz.

Going Further

There are many temperature sensors that work well with Trinket and other Arduino compatible projects. Many sensor programs use floating point math, which is bulky on a Trinket but possible. This includes the TMP36 (Adafruit #165), a popular, low-cost sensor that uses one analog pin. Adafruit released a temperature breakout board in 2014, the MCP9808 (Adafruit #1782), which uses I²C. The Adafruit library to use the 9808 is written for larger Arduinos, but changing the Wire calls to TinyWireM would appear to make it Trinket- and Gemma-compatible. I²C has the bonus that the bus lines may be shared with a display and not take additional data pins.

Running Trinket at 16 MHz

The Trinket 3V will only run at a clock speed of 8 Megahertz (MHz)—operation outside that range is not guaranteed by Adafruit or Atmel. The Trinket 5V is rated to run at a speed of 8 MHz with no code changes, or 16 MHz with a change in code. At 16 MHz, it runs twice as fast, which may be necessary for some critical timing code. More likely, sketches written for the 16 MHz Arduino Uno and similar boards may not run at a different clock rate. The DHT library was written for 16 MHz and the TinyDHT library was left unchanged in that respect, so you need to use 16 MHz with the DHT sensor.

A few code changes are required to have the Trinket 5V run at 16 MHz. First, you need to include the *avr/power.h* file that is standard with the Arduino IDE in the *avr* directory. This defines the value F_CPU. Then you need to add an additional line of code at the very beginning of the setup routine to change the speed using the clock_prescale_set function:

```
#include <avr/power.h>  // needed for 16 MHz on 5v Trinket

// first line in setup sets a 5V Trinket to 16 MHz operation
void setup() {
    if (F_CPU == 16000000) clock_prescale_set(clock_div_1);
    // additional setup code may follow
}
```

No hardware modifications are required and no jumpers need to be set (handy as the Trinket has no hardware jumpers).

For most projects, the default 8 MHz is plenty fast enough (8 million clocks per second!), but if there is a need for speed, the speed is easily doubled with this bit of code. Just ensure the external project circuitry (sensors, displays, time-sensitive components) is expecting such speeds.

Ultrasonic Rangefinding

A Trinket with a display makes an excellent basis for many sensor projects. Pairing a Trinket and a display with a Maxbotix ultrasonic rangefinder, shown in Figure 5-26, creates a handy circuit. Maxbotix makes a wide range of sensors with differing sensing (beam) patterns and sensitivities; these are highly suitable for robotics, proximity sensing, and alarm systems.

The rangefinder reports the distance between the sensor and another solid object. It is commonly used in automobiles and robots for measuring the distance to an object like a wall. You can also use a rangefinder to find distances in rooms or in alarm systems. This project provides a basic distance sensor that you can use for most of these functions.

Figure 5-26. *Maxbotix ultrasonic sensor*

Parts List

- Trinket 5V, Adafruit #1501 or Maker Shed #MKAD69
- I2C character LCD backpack, Adafruit #292
- One monochrome or color LCD display compatible with the LCD backpack, such as the Adafruit #181 monochrome or #398 or #399 color displays
- Maxbotix ultrasonic sensor (LV-EZ1 selected, Adafruit #172)
- Half breadboard, Adafruit #64, Maker Shed #MKKN2, or similar (a full breadboard is fine also)
- Breadboard wires, Adafruit #153, Maker Shed #MKSEEED3, or similar
- 5V power supply, Adafruit #276 or similar
- Female 2.1/5.5 mm DC power connector, Adafruit #368

Build

If you have built the previous Temperature and Humidity project, you can remove the sensor and the wires that connect the sensor to the Trinket. If you have not built the display, refer to the Temperature and Humidity project for assembly and test instructions. When this is done, you can add the ultrasonic sensor per the wiring diagram in Figure 5-27. Figure 5-28 shows the project built on a breadboard.

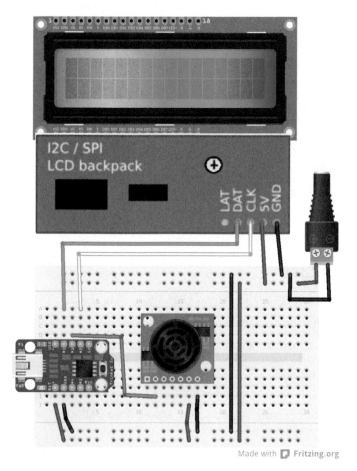

Figure 5-27. *Wiring diagram for the ultrasonic sensor project*

Solder the Adafruit-supplied piece of header onto the sensor so it can be easily plugged into the breadboard. Connect the power and ground wires, along with a wire from the Maxbotix PW pin to Trinket pin #1.

The Maxbotix has a number of ways it can provide data. It can output serial data at the level of its power supply (TTL or V_{CC} level). It can also output an analog voltage proportional to the detection range. Finally, the sensor can send a pulse width signal with a scale factor of 147 microseconds per inch. The first two methods would have been suitable, but it is the pulse width method that turns out to be the simplest to use, translating to the least program space consumed. You'll measure the pulse width with the Arduino `pulseIn` function.

Figure 5-28. *Breadboard view of the ultrasonic sensor on Trinket*

To review the capabilities of this sensor, consult the datasheet on the Maxbotix website (*http://www.maxbotix.com/documents/MB1010_Data sheet.pdf*).

Libraries

You'll use the same TinyWireM and TinyLiquidCrystal libraries you used in the previous project for displaying the distance.

Code

 Change your board type in the Arduino IDE to Trinket 5V 8 MHz before compiling (Tools→Board). If you built the Temperature and Humidity project, the IDE may still be set at Trinket 5V 16 MHz.

The project code is in Example 5-7 and can be downloaded from the repository for this book (*http://bit.ly/GettingStartedWithTrinket*), (directory *Chapter 5 Code*, subdirectory *Chapter5_08Ultrasonic*).

Example 5-7. The Trinket Ultrasonic Rangefinder sketch

```
/* Trinket Ultrasonic Rangefinder Sketch */

#include <TinyWireM.h>              ❶
#include <TinyLiquidCrystal.h>
#define  EZ1pin 1                   ❷

TinyLiquidCrystal lcd(0);    ❸

int8_t arraysize = 9;    ❹
uint16_t rangevalue[] = { 0, 0, 0, 0, 0, 0, 0, 0, 0};
uint16_t modE;           // calculate mode (most common) distance

void setup() {
  pinMode(EZ1pin, INPUT); // set ultrasonic sensor pin as input
  lcd.begin(16, 2);              ❺
  lcd.setBacklight(HIGH);
}

void loop() {
  int16_t pulse;   // number of pulses from sensor
  int i=0;

  while( i < arraysize )
    { pulse = pulseIn(EZ1pin, HIGH); // read in time for pin to transition
      rangevalue[i]=pulse/58;        // pulses to centimeters
                                     // (use 147 for inches)
      if( rangevalue[i] >= 15 && rangevalue[i] < 645 ) i++;    ❻
      delay(10);                     // wait between samples
    }
  isort(rangevalue,arraysize);       // sort samples
  modE = mode(rangevalue,arraysize); // get median, the value desired

  lcd.setCursor( 0,0); lcd.print("Range: ");    ❼
  lcd.setCursor( 7,0); lcd.print("    ");
  lcd.setCursor( 7,0); lcd.print(modE);
  lcd.setCursor(11,0); lcd.print("cm");

  delay(500); // read every half second
}

void isort(uint16_t *a, int8_t n) {    ❽
  for (int i = 1; i < n; ++i) {
    uint16_t j = a[i];
    int k;
    for (k = i - 1; (k >= 0) && (j < a[k]); k--) {
      a[k + 1] = a[k];
    }
    a[k + 1] = j;
  }
}

uint16_t mode(uint16_t *x,int n) {    ❾
```

```
int i = 0;
int count = 0;
int maxCount = 0;
uint16_t mode = 0;
int bimodal;
int prevCount = 0;

while(i<(n-1)) {
  prevCount=count;
  count=0;
  while( x[i]==x[i+1] ) {
    count++;
    i++;
  }
  if( count > prevCount & count > maxCount) {
    mode=x[i];
    maxCount=count;
    bimodal=0;
  }
  if( count == 0 ) {
    i++;
  }
  if( count == maxCount ) { //if the dataset has 2 or more modes
    bimodal=1;
  }
  if( mode==0 || bimodal==1 ) { // return median if there is no mode
    mode=x[(n/2)];
  }
  return mode;
  }
}
```

❶ Include the libraries needed (the Maxbotix does not need a library).

❷ The sensor is on Trinket pin #1.

❸ The display is on the I²C pins, address O (AO–A2 not jumpered).

❹ Define the number of values to calculate the median (sample size, which needs to be an odd number).

❺ Set up the LCD: specify the number of rows and columns, and set the backlight on.

❻ Ensure the value obtained is in range; if so, save the value.

❼ Write the distance to the LCD display via the attached I²C backpack.

❽ This is a sorting function; this code is provided for free use on *http://playground.arduino.cc/Main/MaxSonar* by Bruce Allen and Bill Gentles.

❾ Mode function, returning the mode (most common value) or median (middle value) if it can't determine a mode.

This code compiles to 4,522 bytes of 5,310 available. This leaves almost 800 bytes of code for additional functionality. Using decimal (floating-point) numbers will most likely exceed the program space available.

How It Works

The Arduino IDE function `pulseIn` is perfect for the project, as it returns the pulse width on a pin without requiring much code. This avoids the need for a Serial library or an analog pin. The extra code space afforded by the `pulseIn` function is helpful, as Maxbotix recommends taking several readings and finding the mathematical mode of the data set. The code as written takes nine samples, sorts the values, and finds the mode, which is then displayed on the LCD. The distance is calculated in centimeters by dividing the pulse width by 52 (you can use 147 to obtain the distance in inches).

Troubleshooting

- If you get no display, go to the Hello World I^2C sketch (see "Testing the Display" on page 90) and ensure that the display works.
- If you have no display running Hello World on the I^2C backpack, use the contrast knob to change the LCD display contrast to a readable level. If you decided on an external potentiometer to change contrast and not pin 16 on the backpack, use that.
- If you get no reading of distance, check your wiring from Trinket pin #1 to the PW pin on the Maxbotix sensor and verify that the sensor has its 5V and ground pins connected.
- Make sure you selected Trinket 5V 8 MHz as the board type in the Arduino IDE Tools menu (the temperature sensor sketch used 16 MHz).

Communicating via Serial

In preparation for the final project in the chapter, this section will introduce you to serial communication for Trinket. Serial communication is very useful in providing project status to a connected computer (like your programming computer), or to communicate between projects. It's often used only temporarily, to provide information on what a program is doing while debugging. It is well worth diving into how serial communication can benefit your projects.

The serial port is by far the oldest protocol supported on the ATtiny. It has its roots in the RS-232 standard from 1962. With modern microcontrollers, serial interfaces most often operate at the supply voltage of the processor (often called "TTL levels"), and not the higher voltage levels required of

RS-232. So, the Trinket 5V would operate at 5-volt signal levels, and the Trinket 3V at 3.3 volts.

The serial signals are interfaced with three wires: a send line (transmit, or TX), a receive line (RX), and signal ground. With this wiring, data may be sent and received at the same time. Additional benefits are that serial communication is well understood by the community, uses few data pins, and is implemented on many types of hardware. If you only need to send or receive data, the number of lines can be reduced to two.

A final benefit is that it is easy to interface via serial to larger computers with adapters. One of the most common interface methods is using serial-to-USB chips such as the Adafruit FTDI Friend and similar boards.

The Adafruit FTDI Friend

One specific serial-to-USB interface is the FTDI Friend (Figure 5-29). It uses a popular integrated circuit manufactured by FTDI (*http://www.ftdichip.com/*).

Figure 5-29. *The Adafruit FTDI Friend TTL serial-to-USB board*

You connect the TX designated pin on your Trinket project to the RX pin on the FTDI Friend, and the RX designated pin on your project to the Friend's TX. This crossover conveys the data to the correct pins. You connect the Friend to a computer USB port via a USB male Mini B to Male A cable, as shown in Figure 5-30.

This type of capability also comes in a cable form with the FTDI chip embedded in the USB cable. The functionality is the same. Be aware there are 3.3- and 5-volt signal versions: do not use the 5-volt-only version with a Trinket 3V unless the device is rated for 3.3-volt systems.

Be sure you connect the GND pin on the Trinket to GND on the FTDI Friend. You do not need to connect the Friend's V_{CC} pin to the Trinket circuit. The board draws power from the USB connection.

Talking Serial

Most Arduino code you will see uses the Serial library included with the Arduino IDE. Since the Trinket and its ATtiny85 communicate via methods different from other Arduino compatibles, you cannot use the Serial library. More popular with Trinket is the SoftwareSerial library. Developed for larger Arduino-compatible processors, the library works very well with Trinket. Rather than using USI hardware, the library uses *bit-banging*, toggling digital pins at specific rates to implement a protocol in software. The benefit is that the transmit and receive pins may be any of the five data pins on the Trinket. The same functions found in Serial are in SoftwareSerial, so you can use the code with few changes.

Exploring Serial Use

To become familiar with serial communication, you'll interface an FTDI Friend with a Trinket. A Trinket sketch establishes a serial connection through the FTDI Friend to a computer USB port. This test will power the Trinket from the FTDI Friend to make connections easier. Figure 5-30 shows the circuit.

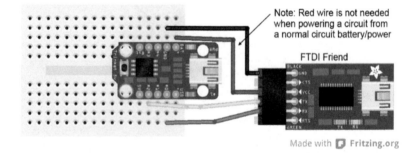

Figure 5-30. *Connecting a serial monitor to Trinket*

Parts List

- Trinket 5V, Adafruit #1501 or Maker Shed #MKAD69
- FTDI Friend, Adafruit #284 or Maker Shed #MKAD22 (alternate is an FTDI cable, Adafruit #70)
- Mini breadboard
- A terminal program on your programming PC
- USB cables

Code

The code for connecting a Trinket to an FTDI Friend is in Example 5-8 and can be downloaded from the repository for this book (*http://bit.ly/Getting StartedWithTrinket*) (directory *Chapter 5 Code*, subdirectory *Chapter5_04Serial*).

Example 5-8. Software serial echo program

```
/* Trinket Software Serial Echo Program */

#include <SoftwareSerial.h> // Use the Arduino IDE library
#include <ctype.h>           ❶

SoftwareSerial Serial(2,0);   ❷

#define LED 1    ❸

void setup(void) {
  Serial.begin(9600);   ❹
  Serial.println("Hello from Trinket"); // Verify comms working
  Serial.println(" ");
  Serial.println("Type text and I will echo it back");
  Serial.println("Any time I see a number, I will flash the LED");
  Serial.println(" ");
  pinMode(LED, OUTPUT);  // Red LED pin
}

void loop(void) {   ❺
  char c;
  if (Serial.available()) {  // Has something been received?
    c = Serial.read();       //  Yes, read it
    Serial.write(toupper(c)); // Write it back out
    if(c >= '0' and c <= '9') flash(); // If a number, flash the LED
  }
}

void flash() {
  digitalWrite(LED, HIGH);      ❻
  delay(500);
  digitalWrite(LED, LOW);
}
```

❶ This is the standard C library for character functions.

❷ Define a software serial connection. The pin order is important: first is the receive pin 2 (abbreviated RX), and second the transmit pin 0 (TX). Both values are required. If you want to only send, use the SendOnlySoftwareSerial library.

❸ The program uses the Trinket onboard LED on pin #1.

❹ Start the serial connection—be sure you set up the computer terminal program for 9,600 baud, 8 bits, 1 stop bit, no parity (a fairly standard set of values).

❺ If you type a character on the computer, it is changed to upper-case and sent back to you. If you type a number, the LED flashes.

❻ This function flashes the LED for half a second as an indicator.

Use

For this project, you will need to download and run a *terminal program* on your programming PC. This type of program provides a blank screen in which serial data can be sent and received. It stems from old terminals used on mainframe computers, but is still very useful today for talking with microcontrollers.

For Windows, PuTTY (*http://www.chiark.greenend.org.uk/~sgtatham/putty/download.html*) is a free program that's widely used. Linux and Mac users may use the Terminal program that comes with the operating system. Set up the communication parameters for 9,600 baud, 8 data bits, 1 stop bit, and no parity. The PuTTy configuration screen is shown in Figure 5-31.

 For Linux and Mac, the operating system also has other terminal programs used with a shell. You can use `screen`, and should specify the port and baud rate, as in `screen /dev/tty.usbmodem 9600`.

You may need to find the serial port used by your programming computer, which you can do by looking at your device list before and after plugging in the FTDI Friend USB cable. In Windows, go to Control Panel→Devices and Printers, right-click the FTDI device, and look on the Hardware tab to see which COM port is being used; then configure PuTTY to use that port.

On Mac or Linux, use the `ls` command to view the list of device files (before and after plugging in the FTDI Friend USB cable) starting with *tty.*, as in:

    ls /dev/tty.*

Limor Fried suggests that for a terminal monitor (which works on any operating system the Arduino IDE works on), you can open another copy of the Arduino IDE in addition to the main one you are using. Select Tools→Serial Port to select the port the FTDI Friend is located on. Then Use Tools→Serial Monitor as your monitor. You might be thinking, "Why not use the current Arduino IDE serial monitor?" The IDE is simple in many respects, port handling included. The IDE window you are using will not work with the Trinket because a Trinket does not have serial onboard, like an Uno. So, another terminal communication program is needed to do the receiving for our use. It seems like a great deal of bother, but after setup, it works rather well.

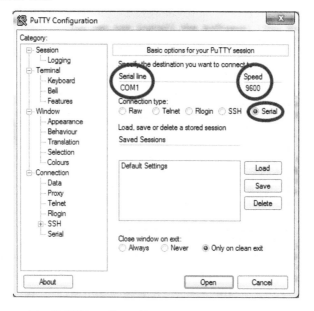

Figure 5-31. *The PuTTY configuration screen*

With a terminal program or monitor, you will get a blank window with a cursor; this is normal.

If the configuration is correct, as you type characters on your keyboard, the Trinket will echo them back. If you type in lowercase, the program converts such characters to uppercase. If you type a number key, the Trinket's onboard red LED will also flash briefly. If you do not get characters, check your software settings to be sure the right serial port and baud rate (9600) are set. Once you get the FTDI Friend and the computer *handshaking* (speaking the agreed-to protocol), things will work well.

In the example code, the SoftwareSerial object was called Serial. Most code for other Arduino compatibles defines the primary serial port as Serial also. Do not think that this is 100% the same. A serial connection is made via SoftwareSerial on the FTDI Friend and not the Trinket USB port.

An external serial connection is helpful for display and debugging. It also allows reuse of example code that assumes the Serial object is present. Debugging via serial can be very helpful. When you're done, you can remove the serial code for final operation to make it run faster with less code space used.

You can use any of the Trinket's five data pins for serial communication (the technical reason is that the ATtiny85 supports *change interrupts* on all data pins). This means the ATtiny can react to changes to its pins and jump to the code that reacts to those changes. Be careful on some pins, though —as the pin diagram in "Connectivity" on page 5 shows, some pins have external hardware that could affect some serial connections. And as always, if you use Trinket pins #3 and/or #4, disconnect the connections to these pins to upload code, then reconnect them afterward. Using Trinket pins #3 and #4 may be helpful when you need pins #0 and #2 for I²C communication, as discussed elsewhere in this chapter.

For your own circuits, you will not power the Trinket from the FTDI Friend. You won't need the red wire in such cases, although the black wire is needed to establish the ground connection.

Other types of communication hardware you may wish to interface to Trinket can also use the serial protocol. This includes many Bluetooth radio modules. You'll learn about Bluetooth in Chapter 6.

Going Further

There are many serial functions available in the Serial library—see *http:// arduino.cc/en/Reference/SoftwareSerial* for a complete list and how they are used.

If you're only sending Trinket data outward, the SendOnlySoftwareSerial library may do the job with less code space used. It is available from author Nick Gammon under a Free Software Foundation license at *http:// gammon.com.au/Arduino/SendOnlySoftwareSerial.zip*. Unzip the code and place the entire code bundle in your Arduino *Libraries* folder, as described in "Installing Libraries" on page 41.

Pulse Width Modulation

Pulse Width Modulation (PWM) is a method used in communication and other circuits. It is so useful that most microcontrollers have hardware to provide PWM signals on some of their pins. On the Arduino Uno, for example, digital pins 3, 5, 6, 9, 10, and 11 all are PWM-capable (marked on the board with a ~ symbol next to the pin number). For the Trinket, pins #0, #1, and #4 can provide hardware PWM.

The Arduino IDE has a function to initiate a PWM signal on a pin, called `analogWrite` (this is a bit of a misnomer, since you're only approximating an analog signal with a square wave on a digital pin). The frequency of the PWM signal on most pins is approximately 490 Hz. Pins 5 and 6 on the Uno (and pins 3 and 11 on the Leonardo) have a frequency of approximately

980 Hz. For the Trinket, PWM via analogWrite works on pins #0 and #1. The call to the function is:

```
analogWrite(pin, duty);
```

where *pin* is the PWM-capable digital pin to write to and **duty** is the duty cycle.

The duty cycle is defined as a value between 0 (always off) and 255 (always on). A value of 128 is 50%, which is half on and half off every period. Figure 5-32 shows various duty cycles for a PWM pulse stream.

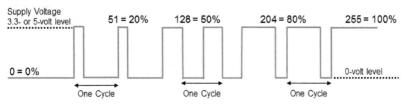

Figure 5-32. *Pulse Width Modulation duty cycles*

The frequency (cycles per second) remains the same, yet at first there is a duty cycle of 0%, then 20% (a value of 51 to analogWrite), 50% (128), 80% (204), and finally 100% (255). The pulse height is the supply voltage (3.3 volts for Trinket 3V, 5 volts for Trinket 5V). You would not normally change the PWM value as rapidly as shown in the figure (from nothing to full), but this makes it easier to visualize.

Pulses through an analog meter can vary the reading on the meter proportional to pulse width. PWM may also be used for dimming LEDs or other uses.

In early versions of the Adafruit-supplied IDE, PWM on Trinket pin #4 could not be initiated by analogWrite, which uses the ATtiny Timer 0 for pins #0 and #1. You may use extra code to set up PWM on pin #4 using Timer 1. The required functions are in Example 5-9 (also available in the repository for this book (*http://bit.ly/GettingStartedWithTrinket*), directory *Chapter 5 Code*, subdirectory *Chapter5_10Pin4PWM*).

Example 5-9. Functions to set up and use PWM on Trinket pin #4

```
void PWM4_init() { // call this function in setup
  // set up PWM on Trinket pin #4 using Timer 1
  TCCR1 = _BV (CS10);                    // no prescaler
  GTCCR = _BV (COM1B1) | _BV (PWM1B);    // clear OC1B on compare
  OCR1B = 127;                           // initialize duty cycle to 50%
  OCR1C = 255;                           // frequency
}
```

```
// function to use an analogWrite-type function for Trinket pin #4
void analogWrite4(uint8_t duty_value) {
  OCR1B = duty_value; // duty may be 0 to 255 (0 to 100%)
}
```

If you use another library that needs the ATtiny Timer 1, this PWM function will overwrite that functionality, and probably cause the other library to not work.

The Analog Meter Clock

Trinket pairs nicely with a real-time clock module to keep time. Using the I²C bus lets you use a real-time clock module along with a display. Once the time is known, you can display it in many different forms. A different take on displaying the time is shown in Figure 5-33. The concept is to display the time using analog panel meters with the scales changed to display the time: one meter for hours, another for minutes.

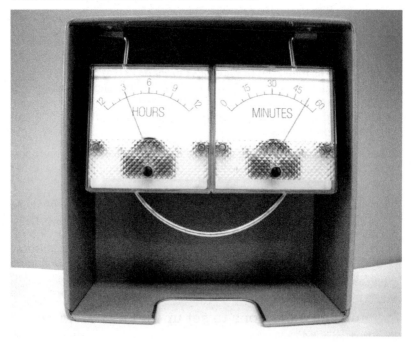

Figure 5-33. *The mounted analog meter clock*

How the clock is packaged is where you get to be creative. The design here uses an open box, lending a modern, floating look. Other designs have been posted on Google+, including a steampunk copper version (*https://plus.google.com/111310175276865788693/posts/ipimrjna9mq*) that is beautiful.

Circuit Design

This project interfaces the Trinket to the Adafruit DS1307 real-time clock (RTC) breakout board. The time is output to two meters that provide readings based on the voltage present on their inputs. The common method for a microcontroller to output a varying voltage is through the use of a digital to analog converter. The Trinket uses another method of providing a varying voltage, via Pulse Width Modulation (PWM) on three of its pins. The meter uses a *moving coil inductance* movement. The coil acts to average the voltage of pulses sent through it. If the pulses are narrow, the average voltage the meter sees is lower, and in proportion the current is lower for the fixed resistance attached to it. For wide pulses, the meter sees nearly the supply voltage (and proportionally full current) and will stay about full scale. This circuit varies the pulse width sent to the meters proportionally to the hour of the day and the minutes after the hour.

For two meters, two of the three Trinket PWM pins are used (pins #1 and #4). The Trinket's third PWM pin (pin #0) is also an I^2C pin required for connection to the clock module, so it is unavailable for display use.

Parts List

- Trinket 5V, Adafruit #1501 or Maker Shed #MKAD69
- Two 50-microamp full-scale analog meters, Adafruit #252
- DS1307 Real Time Clock breakout board kit, Adafruit #264 or Maker Shed #MKAD19
- FTDI Friend, Adafruit #284 or Maker Shed #MKAD22 (optional for debugging)
- Half breadboard, Adafruit #64, Maker Shed #MKKN2, or similar (half is best)
- Breadboard wires, Adafruit #153, Maker Shed #MKSEEED3, or similar
- 5V power supply, Adafruit #276 or similar
- Female 2.1/5.5 mm DC power connector, Adafruit #368
- Perma-proto half-sized breadboard for more permanent installation, Adafruit #1609 or Maker Shed #MKAD49
- Two 100,000-ohm (100K) resistors, 5% or better tolerance preferred
  OR
  Two 92,000-ohm (92K) resistors and two 10K-ohm potentiometers, Adafruit #356

Build

Start by soldering the header pins (provided in the kit by Adafruit) onto the Trinket. The DS1307 kit also requires assembly, as described in the DS1307 Real Time Clock Breakout Board Kit tutorial (*http://learn.adafruit.com/ds1307-real-time-clock-breakout-board-kit/overview*) on building the clock module. This requires some skill with soldering through-hole components.

Wire the project per Figure 5-34. You can power the Trinket with 5 to 16 volts via the BAT+ input and ground. This makes powering the clock very flexible. For this project, I chose the Trinket 5V because the DS1307 board has a 5-volt input that may be connected to the 5V output pin on the Trinket. If another RTC module that works at 3.3 volts were available, the Trinket 3V could be used with appropriate changes to the meter calibration. This clock uses wall power (via a DC power adapter) because the power draw is enough that batteries do not last a long time.

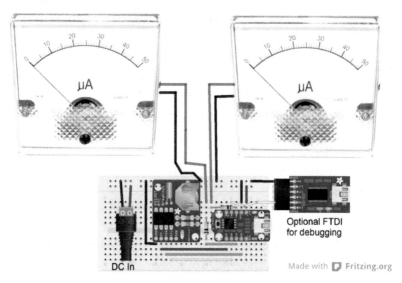

Figure 5-34. *Analog Meter Clock wiring diagram*

All the Trinket pins are used for the permanent circuit, except pin #3 (which is used temporarily to connect to an FTDI Friend). You'll use the SendOnlySoftwareSerial library here to get data from the project to be sure the clock is working during the build. This gives a console-like output using only one pin and ground.

Meters

With a 5-volt Trinket and 50-microamp meters (such as the Adafruit models), for full-scale deflection we need a series resistor on each meter to keep the current less than or equal to the maximum current the meter can handle. Using Ohm's law, $R = V / I$, we get $5 / .00005 = 100,000$ ohms (100K). So, you'll use two 100K resistors, preferably with a 5% or better tolerance. These are commonly available from electronics suppliers. If you want precision in calibrating the meter, you may want to substitute each resistor with a 92K resistor and a potentiometer, nominally 10K ohms, in series. This lets you tune the resistance. When I built the project, the 100K resistors gave accurate enough time without needing potentiometers.

 Do not directly connect the meter to a source of voltage, as it will damage the meter. Use an appropriate series resistor in the circuit to limit the meter current.

Libraries

You'll use the following libraries for this project (you can find the library code at the links listed in "ATtiny-Optimized Libraries" on page 39).

- TinyWireM (the Wire library for the Trinket)
- TinyRTClib (for the DS1307 clock board)
- SendOnlySoftwareSerial (optional for debugging)

Code

Run the sketch in Example 5-10 twice, once to set the clock, and again to have it operate. The code in setup checking for rtc.isrunning() should be uncommented the first time. This will set the clock to the time your code is compiled. You can then comment out that code, since the DS1307 will keep the time with its onboard battery. You can download the code from the repository for this book (*http://bit.ly/GettingStartedWithTrinket*), (directory *Chapter 5 Code*, subdirectory *Chapter5_09AnalogClock*). A separate clock setting sketch is in *Chapter5_11SetClock*.

If you plan to have the code function differently than the sample or have problems you want to debug, a serial monitor will help. The SendOnlySoftwareSerial library is helpful, but not necessary if everything appears to work when assembled. The library adds about 1,300 bytes of program code. You should comment out the serial code in your sketch when it is not needed.

Example 5-10. Source code for the Analog Meter Clock project

```
/* Trinket Analog Meter Clock Sketch  */

#include <TinyWireM.h>
#include <TinyRTClib.h>      ❶
//#include <SendOnlySoftwareSerial.h>

#define HOUR_PIN   1 // Hour   display via PWM on Trinket pin #1
#define MINUTE_PIN 4 // Minute display via PWM on Trinket pin #4

RTC_DS1307 rtc;      ❷
//SendOnlySoftwareSerial Serial(3);      ❸

void setup () {
  pinMode(HOUR_PIN, OUTPUT);      ❹
  pinMode(MINUTE_PIN, OUTPUT);
  PWM4_init();      ❺

  TinyWireM.begin();      // Begin I2C
  rtc.begin();            // Begin DS1307 real-time clock
  //Serial.begin(9600);    // Begin Serial Monitor at 9600 baud
  if (! rtc.isrunning()) {
    //Serial.println("RTC is NOT running!");      ❻
    //rtc.adjust(DateTime(__DATE__, __TIME__));
  }
}

void loop () {
  uint8_t hourvalue, minutevalue;
  uint8_t hourvoltage, minutevoltage;

  DateTime now = rtc.now();      ❼
  hourvalue = now.hour();
  if(hourvalue > 12) hourvalue -= 12; // This clock is 12 hour,
                                      //  convert 13-24 to 1-12
  minutevalue = now.minute();

  hourvoltage = map(hourvalue, 0, 12, 0, 255);      ❽
  minutevoltage = map(minutevalue, 0, 60, 0, 255);

/* Uncomment this and other serial code to check that the clock is working
  Serial.print(now.year(), DEC);     Serial.print('/');
  Serial.print(now.month(), DEC);    Serial.print('/');
  Serial.print(now.day(), DEC);      Serial.print(' ');
  Serial.print(now.hour(), DEC);     Serial.print(':');
  Serial.print(now.minute(), DEC);   Serial.print(':');
  Serial.print(now.second(), DEC);   Serial.print(" - ");
  Serial.print(hourvoltage, DEC);    Serial.print(' ');
  Serial.print(minutevoltage, DEC);  Serial.println();
*/
  analogWrite(HOUR_PIN, hourvoltage);      ❾
  analogWrite4(minutevoltage);
```

```
  delay(5000); // Check time every 5 seconds. You can change this.
}

void PWM4_init() {   ❿
  TCCR1 = _BV (CS10);                  // no prescaler
  GTCCR = _BV (COM1B1) | _BV (PWM1B);  // clear OC1B on compare
  OCR1B = 127;                         // initialize duty cycle to 50%
  OCR1C = 255;                         // frequency
}

void analogWrite4(uint8_t duty_value) {   ⓫
  OCR1B = duty_value; // duty may be 0 to 255 (0 to 100%)
}
```

❶   The TinyRTClib library is an integer version of the Adafruit RTClib Arduino library.

❷   Set up the real-time clock data structure.

❸   For debugging, uncomment the serial code. Use an FTDI Friend with its RX pin connected to Trinket pin #3. You will need a terminal program (such as the freeware PuTTY for Windows) set to the USB port of the FTDI Friend at 9,600 baud.

❹   Define the two meter pins as outputs (these must be PWM-capable pins).

❺   Set Timer 1 to output PWM on Trinket pin #4.

❻   Uncomment these two lines, then run the code once. After that, comment out these lines and load the new code on the Trinket. These two lines set the clock with the date and time from the programming computer when the sketch was compiled.

❼   Get the real-time clock info (hour and minute).

❽   Convert hours and minutes into the pulse width duty cycle for meter display. If you have calibration issues, you can change the last two values (0 higher, 255 lower) to have the needle move less if your scale is not pasted on 100% straight.

❾   The pulse width to the meters is set here and is proportional to the time.

❿   Custom function to initialize PWM on Trinket pin #4.

⓫   Function similar to `analogWrite`, which works on Trinket pin #4.

How It Works

The loop reads the hours and minutes from the DS1307 module. It converts the hours to a 12-hour format, then the hours and minutes are each scaled to a value of 0 to 255 and the pulse widths for pin #1 (hours) and pin #4 (minutes) are adjusted accordingly.

If the serial debug code is enabled, a terminal window will display the code in Figure 5-35.

Figure 5-35. Serial output from the Analog Meter Clock software

The text shows the date and time along with two numbers ranging from 0 to 255, which represent a pulse width corresponding to the time. In Figure 5-35, 255 shows this is noon, and 55 indicates 13 minutes after the hour (55/255, which is not quite 1/4, equates to 13 minutes). If this type of output is not displaying via the serial connection, check the wiring, code, and serial terminal settings.

Preparing Your Meters

You will want to change the meter faces to have them display hours and minutes instead of microamperes. Two basic designs are in Figure 5-36 and Figure 5-37. You can download them from *http://bit.ly/set_up_and_mount_clock*. There are other creative designs on the Internet if you prefer something fancier, including those at *http://bit.ly/custom_meter_backgrounds* and *http://bit.ly/chronulator_support*.

Figure 5-36. The meter face for the hours display

Figure 5-37. The meter face for the minutes display

When designing and printing the meter faces, you can take precise measurements to get the correct needle sweeps, or you can obtain them via some trial and error. If the clock is not giving the correct reading near full scale, the faces reprinting or repositioning the faces instead of looking for a clock or code issue. Use a graphics editing program (Photoshop, GIMP, or many others) to shrink or expand the graphic to fit the same sweep as the original scale on each meter.

If you buy or salvage meters from old electrical equipment, it could lend a different look to the project. The meters may not be 50-microamp full scale, like the modern models, so some trial and error may be needed in finding how much current produces a full-scale reading. Be sure to not send any current through without large series resistors or perhaps a 500K potentiometer. You should use a multimeter to check the resistance.

 Meter movements and needles are very fragile. Be sure when you work on the meter that the needle is not damaged.

To affix the completed faces on the meters, carefully remove the two silver screws on either side of each meter, and lift up the cover. Cut the meter face out of paper. Ensure a semicircle is cut out at the bottom so the needle will swing freely. Use a glue stick or other very light adhesive on the meter face, then carefully slide the new face in without harming the meter needle. Make slight adjustments as needed to align the scale. The meter needle should be pointing at the left-hand mark on the scale. Put the cover back on the meter and screw it on. Using a flat screwdriver, you can adjust the zero on the meter slightly with the black screw in the lower middle of the meter.

Meter Mounting

The meters each have four mounting posts, providing a sturdy mount on nearly any surface. For the author's less-than-traditional mounting method, the meters were placed side by side. The circuit board goes behind the meters. You'll need a flat surface to do this. This is awkward, because the meter movements stick out from the rear. To make a flat area for the circuit, you can fashion a platform, as shown in Figure 5-38, using 3/8"-thick wood precut to 1 3/4" widths. Cut one piece at 2 7/8" to connect the meters, and another at 2 1/4" to bring the level up to the back of the meter movements. You can inset two screws (not too long!) from your screw bin to connect the pieces. The long piece is mounted to the meter with the included nuts. A 3D-printed mount would be another method to make a custom platform.

Figure 5-38. *Mounting meters together and building a circuit board platform*

Mount the circuit board on the back of the meters, as shown in Figure 5-39. Here, the breadboard is mounted. When you are satisfied with the circuit, it would be best to transfer the circuit to a perma-proto or other strip circuit board to provide sturdy, permanent connections.

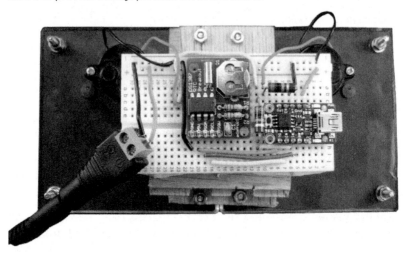

Figure 5-39. *Circuit board attached to meters*

If the meters are mounted in a box, there will be much more flexibility in terms of circuit board placement. You can cut holes for the meter movements and mounting screws. The circuit board mounts anywhere inside the enclosure. Be sure you include a route for power to get to the board.

You can mount the meters in an enclosure of any size or material. A nice metal or wooden box, perhaps? The modern-inspired enclosure in Figure 5-33 is a napkin holder from a clearance sale.

Conclusion

This chapter introduced many concepts through a number of builds: smart LEDs, wearable electronics, movement with servos, sensing and displays, and real-time clocks. With the concepts you now know, you can build a wide variety of projects. In the next chapter, we'll go through some additional builds that demonstrate additional capabilities available in the Trinket.

6/Advanced Projects

The projects in this chapter are a bit more complex. They entail more advanced construction techniques, programming, and interfaces than the projects you've worked on so far. The Trinket, despite being so small, has very advanced functionality built in, which these projects will exploit.

Trinket Jewelry

This project (Figure 6-1), designed by Phillip Burgess, brings Trinket to wearable electronic jewelry. It is based on an 8x8 LED matrix that you can program with user-defined animations. It is a small project, equally suitable as a personal project or one for group workshops. And you will have something eye-catching to wear and show off afterward!

Figure 6-1. *Trinket pendant*

The default animation displays figures from the popular Space Invaders-style games. You can customize the animation—popular themes are Pac-Man-style characters or a beating heart.

Parts List

- Trinket 5V version 1.1, Adafruit #1501 or Maker Shed #MKAD69
  OR
  Trinket 3V Version 1.1, Adafruit #1500 or Maker Shed #MKAD70
  OR
  Adafruit Gemma, Adafruit #1222
- Mini 8x8 LED matrix with I²C backpack, any single color: (Adafruit carries several colors: yellow-green #872, blue #959, yellow #871, red #870, or white #1080)
- 3.7V 150m mAh lithium polymer battery, Adafruit #1317
- LiPo battery charger, Adafruit #1304
- JST surface-mount right-angle connector, Adafruit #1769

Optional parts:

- A lanyard, if you wish to create a necklace. Nonconductive plastic lace (the sort used for weaving bracelets), rubber necklace cord, or heavy fishing line all work.
- A pin backing, for a brooch (Adafruit #1170 or similar).
- A start button (Adafruit #1489 or similar), to activate the animation. Alternatively, you can just use the tiny reset button that is built into the Trinket.
- A bit of heat-shrink tubing (Adafruit #344 or similar) is best for covering some connections; it is cleaner than alternatives such as tape.

Choices

Often-asked questions on parts substitution include:

Can I use a "small" (1.2-inch) LED matrix instead of the "mini" (0.8-inch) version?
Yes, just be extra careful to follow the assembly directions in the Adafruit LED backpack guide (*http://learn.adafruit.com/adafruit-led-backpack/1-2-8x8-matrix*) and install the matrix the right way on the board. The "mini" matrix is recommended for this project because it is more petite and less troublesome to assemble.

Can I use other color matrices?

Yes, you can use any color 8x8 matrices: red, green, blue, white (using the Adafruit models is suggested).

Can I use the Adafruit bicolor matrix backpack?

It will work, but you will have to adjust the bitmaps and code to handle the extra rows. It appears like a 16x8 matrix to the driver chip, but mechanically, the wiring is the same. See *http://bit.ly/bicolor_matrix* for tips.

Tools

This is a soldering project, albeit a small one. You will need a soldering iron, solder, wire (20 to 26 gauge, either stranded or solid), and tools for cutting and stripping wire.

Wiring

Figure 6-2 shows the project built using a Trinket with an external JST connection. The older Trinket 1.0 is shown, but the newer Trinket version 1.1 lets you solder a JST connector on the back to simplify construction. It is best to solder the JST connector on the back of the board before soldering the other connections. Battery power for other uses may be drawn from the BAT+ pin. You can use this for powering other components. For up to 150 milliamps of stable 3.3-volt (Trinket 3V) or 5-volt (Trinket 5V) power, you can use the regulated voltage pin labeled 3V or 5V. Because our display draws more than 150 milliamps, we'll use the BAT+ pin.

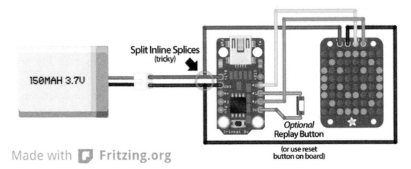

Figure 6-2. *Necklace wiring using Trinket and a LiPo battery*

Figure 6-3 is the same project using the Adafruit Gemma (Trinket's sister board). The Gemma has always had the JST connector but does not have connections for pins #3 or #4.

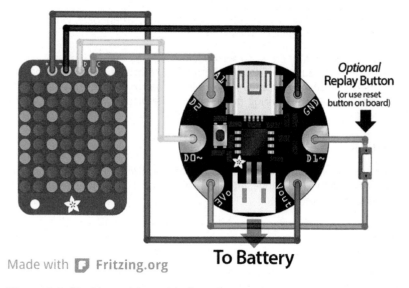

Optional
Replay Button
(or use reset
button on board)

To Battery

Made with Fritzing.org

Figure 6-3. *Necklace wiring using Gemma*

Libraries

Ensure you have the TinyWireM library properly installed in the Arduino IDE. See "ATtiny-Optimized Libraries" on page 39 for library locations and "Installing Libraries" on page 41 for instructions on installing libraries. This project will again use the *power.h* code packaged with the IDE, along with a package called *sleep.h*.

This project also introduces you to including your own code files into the main program. The animation code is defined in a separate file and "brought in" via #include "anim.h".

There is a difference between putting an included file or library (via #include) in angle brackets (< >) versus using double quotes (" "). Both work the same, but by convention standard libraries use angle brackets and local files use quotes. If your code is in the *Libraries* subfolder of your sketchbook folder (*Arduino* for PC/Mac users), you can use angle brackets. The built-in includes are also included with angle brackets (e.g., *avr/power.h* and *ctypes.h*). Local includes should use double quotes to indicate that the file is part of your program package (in the same directory as your *.ino* file).

Code

The main program is in Example 6-1 and is available for download from the repository for this book, (directory *Chapter 6 Code*, subdirectory *Chapter6_01Jewelry*). It uses some new techniques. The animation code is in a separate file listed in Example 6-2 as *anim.h*.

There is enough room in the chip for about 320 frames of animation; anything smaller is fine, of course. *Program memory* (PROGMEM, explained in "Program Memory for Data" on page 134) is used in the *anim.h* file for animations. This is the reason raw I^{2}C command codes are used: to save program space, leaving the maximum room for your animation data.

Example 6-1. Sketch for running the Trinket Jewelry animation

```
// Trinket/Gemma + LED matrix backpack jewelry. Plays animated
// sequence on LED matrix. Press reset button to display again,
// or add optional momentary button between Trinket pin #1 and +V.
// THERE IS NO ANIMATION DATA IN THIS SOURCE FILE, so you should
// rarely need to change anything here. EDIT anim.h INSTEAD.

#define BRIGHTNESS 12   // 0=min, 15=max   ❶
#define I2C_ADDR 0x70   ❷

#include <TinyWireM.h>
#include <avr/power.h>   ❸
#include <avr/sleep.h>
#include "anim.h"    // Animation data is located here

static const uint8_t PROGMEM reorder[] = { // Column-reordering table   ❹
0x00,0x40,0x20,0x60,0x10,0x50,0x30,0x70,0x08,0x48,0x28,0x68,0x18,0x58,0x38,0x78,
0x04,0x44,0x24,0x64,0x14,0x54,0x34,0x74,0x0c,0x4c,0x2c,0x6c,0x1c,0x5c,0x3c,0x7c,
0x02,0x42,0x22,0x62,0x12,0x52,0x32,0x72,0x0a,0x4a,0x2a,0x6a,0x1a,0x5a,0x3a,0x7a,
0x06,0x46,0x26,0x66,0x16,0x56,0x36,0x76,0x0e,0x4e,0x2e,0x6e,0x1e,0x5e,0x3e,0x7e,
0x01,0x41,0x21,0x61,0x11,0x51,0x31,0x71,0x09,0x49,0x29,0x69,0x19,0x59,0x39,0x79,
0x05,0x45,0x25,0x65,0x15,0x55,0x35,0x75,0x0d,0x4d,0x2d,0x6d,0x1d,0x5d,0x3d,0x7d,
0x03,0x43,0x23,0x63,0x13,0x53,0x33,0x73,0x0b,0x4b,0x2b,0x6b,0x1b,0x5b,0x3b,0x7b,
0x07,0x47,0x27,0x67,0x17,0x57,0x37,0x77,0x0f,0x4f,0x2f,0x6f,0x1f,0x5f,0x3f,0x7f,
0x80,0xc0,0xa0,0xe0,0x90,0xd0,0xb0,0xf0,0x88,0xc8,0xa8,0xe8,0x98,0xd8,0xb8,0xf8,
0x84,0xc4,0xa4,0xe4,0x94,0xd4,0xb4,0xf4,0x8c,0xcc,0xac,0xec,0x9c,0xdc,0xbc,0xfc,
0x82,0xc2,0xa2,0xe2,0x92,0xd2,0xb2,0xf2,0x8a,0xca,0xaa,0xea,0x9a,0xda,0xba,0xfa,
0x86,0xc6,0xa6,0xe6,0x96,0xd6,0xb6,0xf6,0x8e,0xce,0xae,0xee,0x9e,0xde,0xbe,0xfe,
0x81,0xc1,0xa1,0xe1,0x91,0xd1,0xb1,0xf1,0x89,0xc9,0xa9,0xe9,0x99,0xd9,0xb9,0xf9,
0x85,0xc5,0xa5,0xe5,0x95,0xd5,0xb5,0xf5,0x8d,0xcd,0xad,0xed,0x9d,0xdd,0xbd,0xfd,
0x83,0xc3,0xa3,0xe3,0x93,0xd3,0xb3,0xf3,0x8b,0xcb,0xab,0xeb,0x9b,0xdb,0xbb,0xfb,
0x87,0xc7,0xa7,0xe7,0x97,0xd7,0xb7,0xf7,0x8f,0xcf,0xaf,0xef,0x9f,0xdf,0xbf,0xff
};

void ledCmd(uint8_t x) {    // Issue command to LED backpack driver   ❺
  TinyWireM.beginTransmission(I2C_ADDR);
  TinyWireM.write(x);
```

```
    TinyWireM.endTransmission();
}

void clear(void) {          // Clear display buffer
  TinyWireM.beginTransmission(I2C_ADDR);
  for(uint8_t i=0; i<17; i++) TinyWireM.write(0);
  TinyWireM.endTransmission();
}

void setup() {                    ❻
  power_timer1_disable();   // Disable unused peripherals
  power_adc_disable();      // to save power
  PCMSK |= _BV(PCINT1);     // Set change mask for pin 1
  TinyWireM.begin();        // I2C init
  clear();                  // Blank display
  ledCmd(0x21);             // Turn on oscillator
  ledCmd(0xE0 | BRIGHTNESS); // Set brightness
  ledCmd(0x81);             // Display on, no blink
}

uint8_t rep = REPS;

void loop() {      ❼
  for(int i=0; i<sizeof(anim); i) { // For each frame...
    TinyWireM.beginTransmission(I2C_ADDR);
    TinyWireM.write(0);             // Start address
    for(uint8_t j=0; j<8; j++) {    // 8 rows...
      TinyWireM.write(pgm_read_byte(&reorder[pgm_read_byte(&anim[i++])]));
      TinyWireM.write(0);
    }
    TinyWireM.endTransmission();
    delay(pgm_read_byte(&anim[i++]) * 10);
  }
  if(!--rep) {              // If last cycle...        ❽
    ledCmd(0x20);           // LED matrix in standby mode
    GIMSK = _BV(PCIE);      // Enable pin change interrupt
    power_all_disable();    // All peripherals off
    set_sleep_mode(SLEEP_MODE_PWR_DOWN);
    sleep_enable();
    sei();                  // Keep interrupts disabled
    sleep_mode();           // Power down CPU (pin #1 will wake)

    // Execution resumes here on wake   ❾
    GIMSK = 0;              // Disable pin change interrupt
    rep = REPS;             // Reset animation counter
    power_timer0_enable();  // Reenable timer
    power_usi_enable();     // Reenable USI
    TinyWireM.begin();      // Reinitialize I2C
    clear();                // Blank display
    ledCmd(0x21);           // Reenable matrix
  }
}

ISR(PCINT0_vect) {} // Button tap interrupt handler (just returns)
```

❶ The brightness of the display is controllable by setting this value from 0 (off) to 15 (brightest). The less bright you set the display, the longer the battery will last.

❷ This is the default address of the Adafruit backpack for the display, unless you change the backpack's A0 or A1 jumpers.

❸ These libraries define functions to put the Trinket into a lower power state.

❹ This table of data is used to realign the rows and columns.

❺ To save program space, you'll define the functions to send a command to the display backpack (ledCmd) and clear it (clear) using low-level calls to the Wire library rather than using a higher-level library.

❻ Some ATtiny85 functions are turned off to save power. The interrupt vector is set to detect the button push, and the display is initialized and cleared.

❼ This code reads the animation data and writes it to the display.

❽ After the last animation frame, the Trinket is put to sleep.

❾ When you press the button, it causes an interrupt, at which time code execution resumes here. All the hardware we need is reenabled, and the animation can start again.

The animation file will come shortly, but first—what is going on in this code?

Normally, when using these matrices (especially with larger Arduino compatibles) Adafruit recommends using their LED Backpack library (*https://github.com/adafruit/Adafruit-LED-Backpack-Library*). As discussed in Chapter 4, libraries can introduce a great deal of code, which limits the amount of space available for your own code. The code in Example 6-1 minimizes the use of external libraries by doing a few things with direct I²C calls—and it may be a bit intimidating at first glance. Here's how it works:

In setup

The code disables the ATtiny85 Timer 1 and all analog-to-digital conversion to save a little power and extend battery life. They are not used by this program. Then it initializes the HT16K33 LED matrix driver chip (using the TinyWireM Wire-compatible library for the I²C protocol), clears the image memory, sets the display brightness, and enables the display (brightness is set with a #define near the top of the code: lower numbers are dimmer and improve battery life).

In loop

The program then loops one or more times, reading animation frames from flash memory (to be explained next), sending the bitmap data to the matrix driver, and displaying each image for a short period. The big table lookup (`reorder`) is needed because the matrix columns are not wired in order on the backpack board. This code reorders the bits in memory to match the column order.

At the end of the sequence, both the LED matrix driver and the CPU are put into a low-power state to help preserve battery life. We then enable a pin-change interrupt on Trinket pin #1 that will wake the CPU from sleep and restart the animation. This button is optional; you can use the on-board reset button as well (though it will have a slight delay because resetting the Trinket means you need to wait for the boot-loader to start).

Animation

The animation data resides in a separate file, so you can modify it without having to rummage through the rest of the code.

At the right side of the Arduino IDE window, click the triangle, select New Tab, and type `anim.h` as the filename. Then load the code from Example 6-2 (also available from the repository for this book, directory *Chapter 6 Code*, subdirectory *Chapter6_01Jewelry*, in the file *anim.h)*.

Example 6-2. Animation data for the jewelry project

```
/* Animation data for Trinket/Gemma + LED matrix backpack jewelry.
   Edit this file to change the animation; it is unlikely you will need
   to edit the source code. */

#define REPS 3      // Number of times to repeat the animation loop (1-255)

const uint8_t PROGMEM anim[] = {    ❶
// Animation bitmaps. Each frame of animation MUST contain
// 8 lines of graphics data (there is no error checking for
// length). Each line should be prefixed with the letter 'B',
// followed by exactly 8 binary digits (0 or 1), no more,
// no less (again, no error checking). '0' represents an
// 'off' pixel, '1' an 'on' pixel. End line with a comma.

B00011000, // This is the first frame for alien #1
B00111100, // If you squint you can kind of see the
B01111110, // image in the 0s and 1s.
B11011011,
B11111111,
B00100100,
B01011010,
B10100101,

// The 9th line (required) is the time to display this frame,
// in 1/100ths of a second (e.g., 100 = 1 sec, 25 = 1/4 sec,
// etc.). Range is 0 (no delay) to 255 (2.55 seconds). If
// longer delays are needed, make duplicate frames.
```

25, // 0.25 seconds

B00011000, // This is the second frame for alien #1
B00111100,
B01111110,
B11011011,
B11111111,
B00100100,
B01011010,
B01000010,
25, // 0.25-second delay

// Frames 3 & 4 for alien #1 are duplicates of frames 1 & 2.
// Rather than list them 'the tall way' again, the lines are merged here...

B00011000, B00111100, B01111110, B11011011, B11111111, B00100100,
B01011010, B10100101, 25,

B00011000, B00111100, B01111110, B11011011, B11111111, B00100100,
B01011010, B01000010, 25,

B00000000, // First frame for alien #2
B00111100,
B01111110,
B11011011,
B11011011,
B01111110,
B00100100,
B11000011,
25, // 0.25 second delay

B00111100, // Second frame for alien #2
B01111110,
B11011011,
B11011011,
B01111110,
B00100100,
B00100100,
B00100100,
25,

// Frames 3 & 4 for alien #2 are duplicates of frames 1 & 2

B00000000, B00111100, B01111110, B11011011, B11011011, B01111110,
B00100100, B11000011, 25,

B00111100, B01111110, B11011011, B11011011, B01111110, B00100100,
B00100100, B00100100, 25,

B00100100, // First frame for alien #3
B00100100,
B01111110,
B11011011,
B11111111,

```
  B11111111,
  B10100101,
  B00100100,
  25,

  B00100100, // Second frame for alien #3
  B10100101,
  B11111111,
  B11011011,
  B11111111,
  B01111110,
  B00100100,
  B01000010,
  25,

  // Frames are duplicated as with prior aliens

  B00100100, B00100100, B01111110, B11011011, B11111111, B11111111,
  B10100101, B00100100, 25,

  B00100100, B10100101, B11111111, B11011011, B11111111, B01111110,
  B00100100, B01000010, 25,

  B00111100, // First frame for alien #4
  B01111110,
  B00110011,
  B01111110,
  B00111100,
  B00000000,
  B00001000,
  B00000000,
  12,          // ~1/8-second delay

  B00111100, // Second frame for alien #4
  B01111110,
  B10011001,
  B01111110,
  B00111100,
  B00000000,
  B00001000,
  B00001000,
  12,

  B00111100, // Third frame for alien #4 (NOT a repeat of frame 1)
  B01111110,
  B11001100,
  B01111110,
  B00111100,
  B00000000,
  B00000000,
  B00001000,
  12,

  B00111100, // Fourth frame for alien #4 (NOT a repeat of frame 2)
  B01111110,
```

```
B01100110,
B01111110,
B00111100,
B00000000,
B00000000,
B00000000,
12,
```

// Frames 5-8 are duplicates of 1-4, lines merged for brevity

```
B00111100, B01111110, B00110011, B01111110, B00111100, B00000000,
B00001000, B00000000, 12,

B00111100, B01111110, B10011001, B01111110, B00111100, B00000000,
B00001000, B00001000, 12,

B00111100, B01111110, B11001100, B01111110, B00111100, B00000000,
B00000000, B00001000, 12,

B00111100, B01111110, B01100110, B01111110, B00111100, B00000000,
B00000000, B00000000, 12,

};
```

❶ The data is coded in the header file, both to make it easy to visualize and easily read by Trinket.

Compile

From the Tools→Board menu, select Adafruit Trinket 8 MHz or Adafruit Gemma, as appropriate. Connect the USB cable between the computer and the board, press the reset button, then click the upload button (right arrow icon) in the Arduino IDE. In a moment, you should get a light show from the LEDs.

 If the display does not light up, check your wiring against the wiring diagram. If the code refuses to compile, most likely the TinyWireM library is not correctly installed, or the *anim.h* file is misnamed. The error messages that appear when you try to compile the code will give you a hint as to which problem might be happening.

You will see an animation sequence of four Space Invaders-style aliens that repeats three times and then shuts off. To see it again, tap the reset button. If the USB cable is still connected, there is a bit more delay due to the boot-loader before it starts again. This is normal: the delay is much shorter when running off the battery or using the optional replay button.

Changing the Animation

To change the animation, you need only edit or replace the contents of *anim.h*. It is rare that you will need to edit the main source code. If you're feeling ambitious, you could write a program to convert an animated GIF into a replacement *anim.h* file, but for now it is necessary to edit this file manually.

There are nine lines for each frame of animation; eight of these are bitmap data, and the ninth line is the delay time. Each bitmap line consists of the letter B followed by 8 binary digits (0 or 1), where 0 (zero, not uppercase letter o) represents an "off" pixel and 1 (one) an "on" pixel, and ends with a comma. The delay is given in 1/100ths of a second; 100 = 1 second, 25 = 1/4 second, and so forth. The delay range is from 0 to 255; if you need longer delays, make duplicate frames.

You can almost see the bitmap image in Figure 6-4.

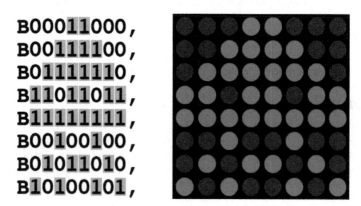

Figure 6-4. *Comparing the code and the LED matrix*

You do not have to represent every row on its own line like this, but it makes the image much easier to visualize. You can see a few places in the code where all nine lines are placed together to save space in the vertical direction. These frames are copies of others; you already know what they look like so you don't need to see them laid out nicely.

After editing, press reset on the board and upload as you did before.

If the program refuses to compile after editing *anim.h*, the cause is most likely one of the following:

- A missing comma at the end of a line
- A missing uppercase B at the start of a line

- Too many or too few digits on a line, or characters other than 0 (zero) and 1 (one)
- Spaces between characters

Finishing the Jewelry

Connect the LiPo battery to make sure that everything runs, then stack the components and fold any wires around so that nothing is protruding. If you're using the on-board reset button, make sure it is on the back side where you can reach it, not blocked by the LED backpack or battery. Other than your wires, there should be no conductive parts making contact between the microcontroller board and LED backpack.

Figure 6-5 shows building a pendant by mounting the Trinket or Gemma on the back of the display. The layers can be held together with small pieces of foam tape, dabs of hot-melt adhesive, or epoxy. If you added a replay button, you can encase the whole thing in a plastic bubble (such as the ones toy vending machines dispense) with just the button protruding.

Figure 6-5. *A side view of the pendant built with a Gemma*

The mounting holes on the LED backpack can be used for attaching a lanyard. Plastic lace from a craft store is one option. Do not use a metal chain for this, as it is conductive and could cause an electrical short as the pendant shifts around. Another option is a pin back. Adafruit has a magnetic version, but you can find the regular pointy-pin variety at most sewing or craft shops. Be careful to make sure the pin doesn't bridge any electrical contacts.

The little battery should last all day, depending on how often you activate it. To recharge it, unplug the LiPo cell from the Gemma or Trinket and connect it to the Adafruit micro LiPo USB charger (or equivalent). The battery does not charge when the microcontroller is connected to USB.

Program Memory for Data

The Trinket's RAM is so very precious: only 512 bytes. Where can you stash additional data? The EEPROM is available, but it's not easy to use and is also limited to 512 bytes. It is more practical to use program (flash) memory to store additional data.

You can use program memory in a couple of ways. When declaring variables, you can declare them as constant data, as is done in the jewelry program:

```
#include <avr/progmem.h>

const variabletype PROGMEM variable = { data; ... }
```

The PROGMEM identifier states that the constant data you define should be placed into flash memory and not into RAM. The types of data you can use for PROGMEM are defined in the standard header file *avr/progmem.h*, which is included with the Arduino IDE.

The problem with this method is that the data cannot be read back with normal data functions. Some cryptic bugs are generated by using ordinary data types for program memory access! You must read the data back using calls to function macros prefixed by pgm_read, as in the following examples:

```
// read back a two-byte integer
_myInt_ = pgm_read_word(_wordprogmemvariable_)

// read back a character (single byte)
_myChar_ = pgm_read_byte(_charstringprogmemvariable_);
```

You used pgm_read functions in the loop function in Example 6-1.

The best programming practice is to read small chunks—such as one frame of animation data—then use it, then repeat. It would make little sense to define a large block of data as PROGMEM and then read all of it into RAM.

Program memory can also be useful for storing strings. We do this by enclosing them in the F macro. Consider this example:

```
Serial.println("Long string we want to send to a SoftwareSerial pin");
```

Storing this string (the characters, and a null character to terminate the string in memory) uses up to 62 bytes of RAM. We can shift this to program memory as follows:

```
Serial.println(F("Long string we want to send to a SoftwareSerial pin"));
```

The F macro tells the compiler to place this string into flash memory. Its use remains unchanged, but the program will have more RAM available.

The F macro does add some bulk to the compiled code—about a hundred bytes are required to implement the functionality. If you want to use the macro to save space in a very constrained program memory environment, the space saved may be less than the space used to implement the macro.

More on use of program memory can be found on the Arduino website (*http://arduino.cc/en/Reference/PROGMEM*).

Trinket Occupancy Display

This is a fun project inspired by an Adafruit customer email inquiry.

Every facility has a conference room or meeting space. And when the door is closed, it is always a guessing game whether the room is occupied or not. This inevitably leads to someone opening the door and disturbing what is happening inside—interrupting a meeting or perhaps spoiling an important experiment. An occupancy indicator, such as the one in Figure 6-6, could solve the problem.

Figure 6-6. *The Trinket Occupancy Display*

A passive infrared sensor (PIR) is the gold standard for tracking movement of people in a general area. It does not measure distance as an ultrasonic sensor does, but it has a wide field of view and good sensitivity to warm bodies. You can find a tutorial on this type of sensor at *http://bit.ly/PIR_Motion_Sensor*.

In the project we'll build next, a Trinket receives the signal from a PIR inside a room. When the room is occupied, the sensor activates and the Trinket will display a red X on an Adafruit 8x8 bicolor LED matrix. If there is no movement, it displays a green square.

Commercial sensor/indicator combinations such as this can cost over $400 dollars each. If you have many rooms, the cost can add up.

 The occupancy indicator may be built on a breadboard if you do not intend on making a permanent project.

Parts List

- Trinket 5V, Adafruit #1501 or Maker Shed #MKAD69
- 8x8 bicolor LED matrix with I^2C backpack, Adafruit #902
- PIR (motion) sensor, Adafruit #189
- Perma-proto half-size breadboard (single), Adafruit #1609 or Maker Shed #MKAD49
- 5V power supply, Adafruit #276 or similar
- Panel mount 2.1/5.5 mm DC power connector, Adafruit #610
- Small plastic box, weatherproof, Adafruit #903
- Cable gland, PG-9, Adafruit #761
- Female header, Adafruit #598 or similar
- Male header, Adafruit #392 or similar
- Rainbow female–female cable, Adafruit #793 or similar
- 1,500-ohm (1.5K) resistor
- Four-conductor wire, long enough to run from your PIR to your box
- 8-32 × 1/2" screw

Tools

- Soldering iron and solder
- Drill and bits (a rotary tool may substitute for some work)
- File(s)
- Diagonal pliers for cutting wires and header
- Wire strippers
- Chisel and hammer

Wiring

The half-size perma-proto circuit board is nearly the perfect size to fit the planned enclosure.

Cut some female headers into two lengths of five and two lengths of four pins. The fives will make a socket for the Trinket, allowing easy access to remove it for programming. The four-pin blocks are for the bicolor LED display signals (bottom) and to balance it on the other side, as shown in Figure 6-7 and Figure 6-8.

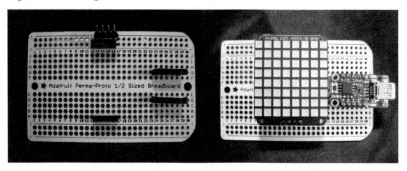

Figure 6-7. *Attaching headers to mount the Trinket and the LED matrix*

Solder the wires and the header in place. A bit of Blu Tack or other temporary adhesive is helpful to keep parts in place when you turn the board over to solder. Insert the 1.5K resistor and solder. Trim off the excess leads. The male header pins allow for easy removal of wires for the PIR and power (if you used female headers for the power jack).

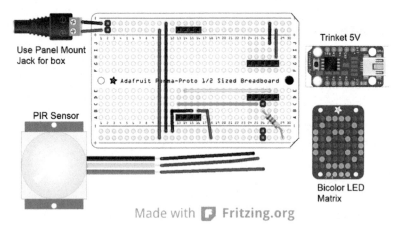

Figure 6-8. *Wiring for the Trinket Occupancy Display and sensor*

Now mounting the large parts is as easy as snapping them on, as shown in Figure 6-9.

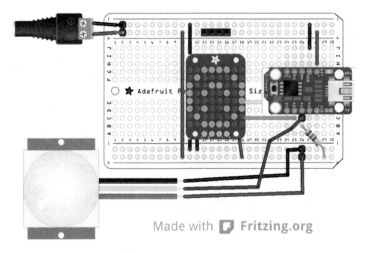

Figure 6-9. *Parts mounted to the board—note the actual display is larger than the Fritzing library shows*

The display top will be supported by the top piece of a four-pin header (which does nothing electrically and has no pins to plug into; it is only a brace). Figure 6-7 shows a small piece of bent header obtained from the scrap pile to prop it up a tiny bit more—you can use nearly anything, such as Play-Doh, hot glue drops, etc.

Make the point-to-point wiring connections as follows:

- Trinket pin #0 to the display I^{2}C data line
- Trinket pin #2 to the display I^{2}C clock line
- BAT+ on Trinket to the + line on the proto-board, which is connected to 5 volts
- GND on Trinket to the − line on the proto-board, which is connected to power ground
- Trinket pin #1 to a 1,500-ohm resistor, which is connected to +5 volts (resistor available at RadioShack, Maker Shed, and other electronics outlets)
- Trinket pin #1 to the PIR data line (center)

The red power line on the PIR and display goes to the 5-volt + line on the proto-board. The black ground line on the PIR and display goes to the ground line on the proto-board.

Be sure you interconnect the top and bottom power lines with wires (toward the left).

Mount two pins to the power lines to plug in the power jack (the upper left of the circuit board). You'll also mount pins on power and Trinket pin #1 to easily connect the PIR wire. For a final install, you will probably connect a long four-conductor wire from the box out to the PIR mounting location.

Double-check all wiring with the diagrams and pictures.

Libraries

You'll use three Arduino libraries to facilitate programming:

- The TinyWireM library provides I^{2}C communications between the Trinket and the display
- The Adafruit-LED Backpack library has routines to talk to the display
- The Adafruit GFX library is required to accompany the backpack library

See "ATtiny-Optimized Libraries" on page 39 for library locations and "Installing Libraries" on page 41 for instructions on installing libraries. The Adafruit libraries take up a fair amount of space but simplify programming.

Code

The libraries make the code compact, as you can see in Example 6-3 (also available from the repository for this book, directory *Chapter 6 Code*, subdirectory *Chapter6_02Occupancy*). Two bitmaps are defined in program memory, similar to the coding method in the Trinket Jewelry project.

Example 6-3. Sketch for the Trinket Occupancy Display project

```
/* Adafruit Trinket-based Room Occupancy Sensor and Display */

#include <TinyWireM.h>
#include "Adafruit_LEDBackpack.h"
#include "Adafruit_GFX.h"              ❶

const int PIRpin = 1;      ❷
uint8_t pirState = LOW; // Stores state of the PIR sensor

Adafruit_BicolorMatrix matrix = Adafruit_BicolorMatrix();

void setup() {
  pinMode(PIRpin, INPUT); // Initial state is low
  matrix.begin(0x70);     // Pass in the address
}

static const uint8_t PROGMEM // X and square bitmaps     ❸
  x_bmp[] =
{ B10000001,
  B01000010,
  B00100100,
  B00011000,
  B00011000,
  B00100100,
  B01000010,
  B10000001 },

  box_bmp[] =
{ B11111111,
  B10000001,
  B10000001,
  B10000001,
  B10000001,
  B10000001,
  B10000001,
  B11111111 };

void loop() {
  int sense = digitalRead(PIRpin); // Read PIR sensor
  if(sense == HIGH) {            // If high and it was low, sensor tripped
    if(pirState == LOW){         //   and we display a red X
      matrix.clear();
      matrix.drawBitmap(0, 0, x_bmp, 8, 8, LED_RED);
      matrix.writeDisplay();
      pirState = HIGH;
    }
  } else {
```

```
if(pirState == HIGH) {        // If low and state was high, sensor set
    matrix.clear();           // and we display a green box
    matrix.drawBitmap(0, 0, box_bmp, 8, 8, LED_GREEN);
    matrix.writeDisplay();
    pirState = LOW;
  }
 }
}
```

❶ The Adafruit graphics library adds a great deal of unneeded code, but simplifies this project significantly. Using "raw" Wire library calls as in "Trinket Jewelry" on page 121 could result in less code.

❷ The PIR signal pin is connected to Trinket pin #1.

❸ These bitmaps may be changed to any design you desire.

The PIR sensor is read from Trinket pin #1. Because this pin has the onboard red LED, a low-value pull-up resistor (1,500 ohms or so) is required. The pin is set to an input and the pin state is read in a loop. When a state change is detected (i.e., the PIR senses a change in the state from the last time through the loop), the bitmap changes. A red X is displayed for occupied, and a green square for unoccupied.

 The Arduino **delay** software function cannot be used with the libraries, as it adds a bit too much code (the program uses nearly all the Trinket program space). Adjust time delays using the PIR potentiometers.

Unfortunately, this much code takes most of the program flash memory. If you add much more code to this program, it will most likely give an error indicating you've used too much program memory. The code space is very tight due to the nice Adafruit prewritten libraries.

Enclosure and Board

The small weatherproof enclosure (Adafruit #903) is a good-size box for this project. The half-size perma-proto board fits inside the box and clears the rounded divots with a bit of modification.

You should modify the board and box as shown in Figure 6-10.

Round the edges on one side of the perma-proto board with a file or Dremel-style rotary tool. Enlarge the hole in the center line on the side with the rounded edges to fit an 8-32 1/2" screw that mates with the box.

You'll use the 2.1 mm panel mount barrel jack for power. *Solder the power wires on prior to installation* (the large lug is positive). Colored wires with female pin connectors on them would be best, if you have some; otherwise, long stranded wire will work well. Connect the red wire to the center lug, and the black wire to the lug 90 degrees clockwise from that. Connect the power to the jack and use your multimeter on a voltage scale to test for correct voltage and polarity. It is best to test the voltages before connecting the circuit board to avoid burning out a part.

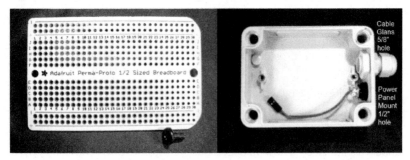

Figure 6-10. *The modified circuit board and mounting box*

Drill a 1/2-inch hole in the box just past the cover hole. Chisel out the plastic dimple in that area to get the power jack to fit in that location. Thread the jack in, securing it with the included plastic screw ring. The cable gland keeps things weathertight where you run the PIR sensor wires through the enclosure. Drill a 5/8-inch hole (for the large gland; smaller if you use the small gland) between the power jack and the other case cover hole. There is a small lip on the case: grind it down a bit with the rotary tool to have the gland lock ring fit snugly. You can run penetrations through the back as an alternative if you want the look to be "cable free," as for a nice conference area.

It is best to make mounting holes in the back of the case if needed at this point. They will let you secure the case on a wall. This would be harder to do later in the build, given the number of items being mounted in the box.

The circuit board should be mounted so the screw is in the standoff furthest from the penetrations. The other side rests on the other mount but is not quite long enough to mount via the hole. This is fine, as that side of the enclosure is crowded with box penetrations.

Box Connections

Mount the circuit board inside the enclosure with the screw. If it does not fit, you must take some material off the right ends and enlarge the hole for the screw.

Plug in the power from the power panel mount jack to the plus and minus on the board. Triple-check the power connections. Run a cable from the PIR to the +, −, and pin #1 on the Trinket, per Figure 6-8. You can now plug in the display.

Adjustment

The PIR has two adjustment potentiometers on the back, as shown in Figure 6-11.

Figure 6-11. *Available adjustments on the Adafruit passive infrared sensor board*

The one potentiometer adjusts the sensitivity of the sensor. Start with a reading toward the "min" side, and adjust clockwise as necessary.

The other potentiometer adjusts the time the sensor stays *latched*, or in the on state. When testing, leave this at the left, which is a short time. Adjust clockwise for longer intervals.

Test the unit by making all the connections, with the PIR pointed away from movement or covered by a cardboard box. The green square should be displayed, as in Figure 6-12. Figure 6-13 shows what it looks like when motion is detected.

Figure 6-12. *Testing the project: no motion detected*

Figure 6-13. *Testing the project: motion detected*

Move in front of the sensor and the red X should be displayed, as in Figure 6-13.

If the display does not change, check your wiring to the sensor and the PIR sensor adjustment potentiometers. Once it is working, you can again adjust the potentiometers to increase the sensitivity, the delay, or both.

Room Placement

Mount the display box above or next to the door of the room you want to monitor. Inside the room, mount the PIR sensor so it has a wide field of view , as in Figure 6-14. For a conference room, aim for the table area, taking care not to point above people's heads. Five-volt power should be obtained from a wall adapter and connected to the display. Run a wire from the display box to the infrared sensor.

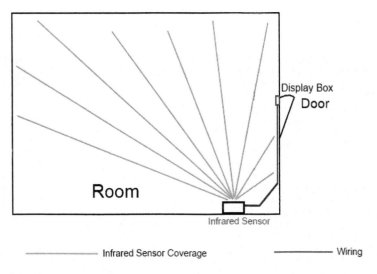

Figure 6-14. *Mounting the sensor and the display box*

If you want just the display to show through the clear cover, you can make a mask from white paper or vellum. You can have the LED display show through the mask or cut out a square just for the display.

Going Further

The basic function of "sense and warn" applies to a broad range of useful projects. You can replace the sensor used here with other models. You can also replace the display with a character display or NeoPixels.

A full-featured alarm system is well within the Trinket's capabilities. The next project is a more standard alarm that can be used to monitor rooms and alert you wirelessly.

Trinket Alarm System

A sensor system that acts as an alarm is something nearly every experimenter tries to build at some point. The leap from a simple sense and warn circuit to a practical alarm system can be more complex than it appears at first glance, though. "Bad guys" have been defeating alarms since the first use of dogs and geese as perimeter alarms. Bad guys know you have an alarm, and may possibly know how to get around it—that is, unless you have put thought into the alarm, staying ahead in the "what if" game.

Building on previous projects, you can fashion an alarm system like the one shown in Figure 6-15, which is suitable for one or more rooms. The scenario is a common one: an unprotected room, with multiple ways to break in. Your Trinket alarm will transmit some type of signal alerting you to any activity.

Figure 6-15. *A Trinket-based alarm system*

Parts List

- Trinket 3V, Adafruit #1500 or Maker Shed #MKAD70
- Bluefruit EZ-Link serial breakout, Adafruit #1588
- PIR (motion) sensor, Adafruit #189
- Magnetic contacts SPST, normally closed, Adafruit #375
- Perma-proto quarter-sized breadboard (single), Adafruit #1608 or Maker Shed #MKAD48
- 5V power supply, Adafruit #276 or similar
- Panel-mount 2.1/5.5 mm DC power connector, Adafruit #610
- USB LiPo charger, Adafruit #259
- 3.7V 1,200 mAh lithium polymer battery, Adafruit #258
- Terminal block, four-pin Eurostyle, Adafruit #677
- Small plastic box, weatherproof, Adafruit #903
- Cable gland, PG-9, Adafruit #761
- Female header, Adafruit #598 or similar
- Resistors: 1,000 ohm, 1,500 ohm, 2,200 ohm, 3,300 ohm
- Hookup wire, Adafruit #289, #288, #290, or similar
- Two-conductor wire, to run from the magnetic sensors to the box
- 8-32 × 1/2" screw
- Optional: Rainbow female–female cable, Adafruit #793 or similar
- Optional: Plastic standoff, double-sided foam tape

Tools

- Soldering iron and solder
- Drill and bits (a rotary tool may substitute for some work)
- Files
- Diagonal pliers for cutting wires and header
- Wire strippers
- Chisel and hammer

Theory

Most alarm systems follow a basic design: a switch closure (or opening) triggers action by a central monitor, and the alarm generates some form of annunciation. A block diagram is in Figure 6-16.

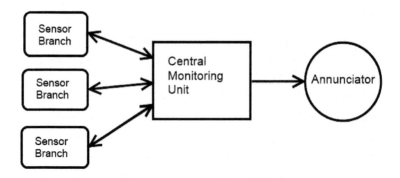

Figure 6-16. *The building blocks for an alarm system*

The sensor blocks may actually be multiple sensors in one monitoring branch; all of the sensors in the branch are connected into the alarm as one bundle. The system might be designed to give only one indication that the sensor chain has been triggered, or, with a smarter design, multiple sensors can provide individual indications.

Switches can be normally closed or normally open, as shown in Figure 6-17. When they are triggered, they change state (closed goes open, open goes closed). The monitoring unit notices the change and, if it meets criteria that the monitor believes is an alarm, the monitor annunciates.

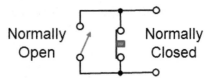

Figure 6-17. *Switch types*

Annunciations can take many forms. The stereotypical alarm system has a large horn to make an ear-splitting sound, but it can be much subtler. The central monitor can produce a "silent alarm," making the intrusion known in some way, either locally or far away. An example would be a text or SMS message on your phone stating when an alarm has been tripped. The simplest alarm may only trigger a local alarm not designed to scare the intruder—this can be a nearby notification, or it could be a record of alarm events for later review.

Multiple Sensors, One Pin

You can use nearly unlimited sensor switches if you place multiple switches in each branch as shown in Figure 6-18. In a simple design, the branch can sense when one switch out of the group changes state, but all you know is that one of the multiple switches in the branch has tripped. Construct a branch by using normally open sensors in parallel or normally closed sensors in series. You can use both types of sensors, putting the closed switches in line and the open switches in parallel. The pull-up resistor pulls the circuit high if the loop is opened; otherwise, the loop is grounded.

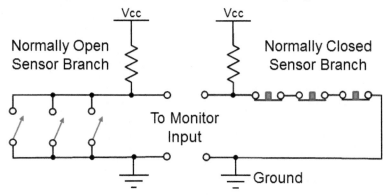

Figure 6-18. *Normally closed and normally open sensor branches*

Alternatively, there are circuit methods that allow you to identify which switch in a branch was tripped if you use an analog pin. These methods use resistors to change the voltage values that the analog pin reads. Some Arduino shields use this method to read four to six switches, to determine which button is pushed.

However, many analog switch reading circuits have a problem: they cannot determine if two buttons are pressed at the same time. For a simple alarm system, this might not matter—you will receive notification that a sensor indicates an alarm, but you won't know which specific sensor activated, only that one sensor on a particular branch has triggered. With a bit more circuit design, though, you can determine if multiple sensors have tripped on a single branch, and which ones they were. The method used most often in textbooks is the R-2R resistor ladder shown in Figure 6-19.

Figure 6-20 shows a simplified design by the author. It is a parallel resistor system that uses fewer components and has good accuracy for Arduino-type analog inputs. The complexity grows with the number of sensors, so it was designed for just three sensors on a branch.

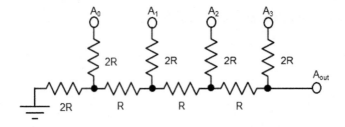

Figure 6-19. *R-2R ladder*

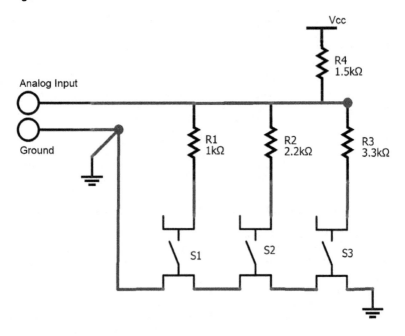

Figure 6-20. *Sensing three switches using one analog pin*

Normally, the analog input is pulled high by resistor R4. If any of the switches S1, S2, or S3 is closed, the resistance changes in a predetermined way. Circuit-wise, each of the resistors R1, R2, and R3 would add to the total resistance, using the familiar circuit formula 1/R = 1/R1 + 1/R2 + 1/R3. All you have to do is measure the analog values read by the Trinket and add them to the code. This method works equally well for normally open or normally closed switches. It requires four resistors for three sensors, whereas the pure R-2R ladder method requires five to six resistors.

Project Design

This design will take sensors, add central monitoring code, and output alarm events to an annunciation system (another computer connected via Bluetooth). Several configurations will be shown to allow you to configure different alarm systems.

The Trinket has the capability to create an excellent alarm at a cost much lower than that of other alarm platforms. There are five general-purpose pins, three of which may be digital or analog. At a minimum, you will need one sensor (or sensor branch) in, and one annunciation path out. That uses two of the five pins. Or you can have an annunciator that uses up to four pins, or several sensor branches and a more modest annunciator.

This project design is for a generic monitored area: a single area alarmed and annunciated either locally or to a remote site. You can adapt the design for multiple rooms (doors, windows, movement), but there are limitations to the number of sensors if you want to know the exact sensor that is triggered. If you only need to know whether "something" happened in a sensor chain, you can use a nearly unlimited number of sensors.

A typical scenario is monitoring a room in a building or a garage, as shown in Figure 6-21. In this garage example, we want to monitor the main door, a side door, and the space within the room (in a regular room, we might substitute a window for the garage door). To cover the area well, we'll use a magnetic contact on each of the areas that can open and close, plus a volumetric (area) sensor in case one of the other sensors fails or someone breaks in without tripping a magnetic contact. You may go to greater extremes in covering an area, but it costs resources (money and processor pins) without providing much more security.

This design uses one analog sensor branch, shown in the design diagram. The infrared sensor (PIR) and the magnetic contacts are wired together into an analog ladder, as in "Multiple Sensors, One Pin" on page 149. Annunciation is transmitted via Bluetooth to a monitor room inside the house. In a good design the monitor is located away from the area being surveilled, so the intruder cannot disable the alarm easily.

This design also leaves pins available for additional sensors or other annunciation.

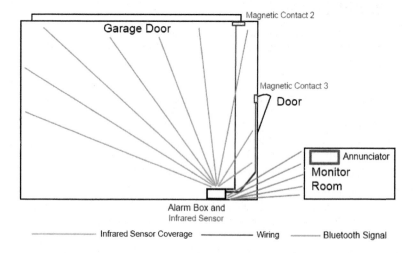

Figure 6-21. *Sensor layout for an alarmed space and monitor location*

Annunciation Selections

There are many ways to alert you if the alarm has tripped. Some favorite ways:

A speaker

The simplest method is to use a loud noise. A piezo speaker does this well; you can output several tones or just an ear-piercing whine. You can use an audio amplifier to make a louder sound. You will need to match the wattage of your speaker to that of the amplifier. Between the Trinket and the amplifier, ensure you do not input too much power or voltage.

A display

If you leave pins #0 and #2 (the I²C pins) free, a character display with a backpack, as demonstrated in previous projects, could display the status. You may also use an LED as a simple indicator. The Trinket pin #1 built-in LED is perfect for this, although you could use NeoPixels or other LEDs.

Communication to other devices

A serial connection may also be established using an Adafruit FTDI Friend or Bluefruit EZ-link. Another computer or handheld device may receive the serial data and act on it.

A combination

If the infrared sensor and all the magnetic switches are on a combined branch on an analog input, then multiple annunciation methods could be available.

The final project build plan annunciates via both the pin #1 LED and a Bluefruit.

Build

This project uses the exact same enclosure and similar construction methods (Figure 6-22) to those used in "Trinket Occupancy Display" on page 135.

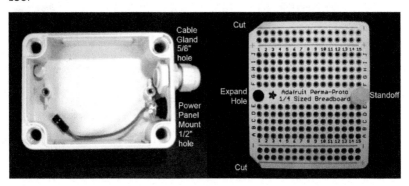

Figure 6-22. *The alarm mounting box and circuit board modifications*

Adafruit's small weatherproof enclosure is economical for this project, although the layout will be a bit tight (things are layered in the enclosure). If you need additional room, a larger enclosure can be substituted. The quarter-sized perma-proto board is a great size to lay out circuit components.

The 2.1 mm panel mount barrel jack is used for power: solder power wires on prior to installation (the large lug is positive). Drill a 1/2-inch hole just past the cover hole. You'll need to chisel out a pesky plastic dimple to fit the jack in that location. Thread the jack in, securing it with the included plastic screw ring. Use the large cable gland to run sensor and signal wires through the enclosure and keep things weathertight. Drill a 5/8-inch hole between the power jack and the other case cover hole. There is a small lip on the case: Dremel or shave it down a bit to have the gland fit snugly against the outer side.

It is best to make mounting holes in the back of the case now, if they'll be needed to screw the case into a wall. Doing this later in the build would probably be harder, given the amount of items you'll place inside the box.

Only small modifications are needed for the circuit board. The corners should be cut just a bit and the left hole expanded slightly to allow the board to fit snugly when screwed on one brass mounting hole. Fit a small 8-32 screw on the right, and optionally place a standoff through the left hole.

Populating the Board

Figure 6-23 shows the parts placement.

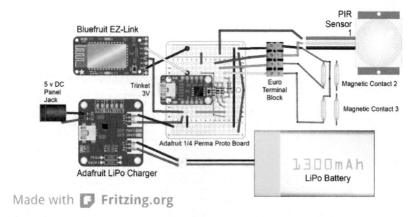

Figure 6-23. *Wiring diagram for the Trinket alarm system*

Wiring proceeds as follows:

1. Use two pieces of female header, five pins each, to mount the Trinket to the board. As this project uses Trinket pin #3, you'll need to be able to easily take the Trinket off the circuit board for programming. See "Trinket Occupancy Display" on page 135 for how the headers are mounted.

2. Place single male pins on the negative (ground) power line, the Trinket BAT+ pin, and Trinket pin #0; you'll use these to connect the Bluefruit later. You can also use hookup wire connected directly to the circuit board, but the wires will not be easy to disconnect if there are issues.

3. Four resistors provide the analog divider to multiplex three sensors. The 1,500-ohm resistor (the large one in Figure 6-24; your resistor may be smaller) goes between the Trinket 3.3-volt regulated pin and the common resistor junction. Each of the other resistors is connected to this common area and to Trinket pin #3 (which is also analog pin 3). Four wires are run off-board to connect the sensors via the Eurostyle terminal block.

4. The black wire connects the ground, and the red wire connects the power. The board receives power via a JST connector with attached wires that comes with the Adafruit LiPo charger board (near the top of the photos). The red wire goes to the Trinket BAT+ pin, and the black to the pin labeled GND. Other parts of the circuit draw regulated power from the pin marked 3V (which is regulated by the Trinket to 3.3 volts and provides up to 150 milliamps).

5. The connector for the PIR (infrared sensor) is also connected to regulated power (bottom of Figure 6-24).

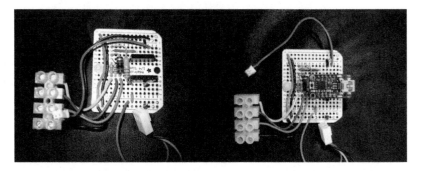

Figure 6-24. *Wiring the alarm circuit board and connectors*

Double-check your connections, or even triple-check them!

Place the battery in the case as shown in Figure 6-25. The LiPo charger board goes on top, near the power connector. Double-sided foam tape may help here. Solder or plug in the wires from the power connector to the DC IN connections on the charger (near the USB connector). The female–male connectors on the DC IN on the LiPo board make it easier to connect and disconnect, but you must carefully observe the polarity every time you make the connections. If the project ever fails to light up, check these wires.

At this point, you should plug in the 5-volt power supply and charge the LiPo battery.

Screw the wired circuit board into the threaded connector with an 8-32 1/2" screw (not provided with the enclosure) or otherwise secure it as shown in Figure 6-26. A plastic standoff or other resting point in the opposite mounting hole area is a good idea. I used the remaining threads of the standoff above the nut to secure another standoff for the PIR later.

Wire the three sensor wires to a terminal strip. The strip allows wires to be connected and disconnected as the system is installed. The connections are, from the bottom: the common (ground) connection, then the PIR switch wire (yellow), then magnetic sensor #2 and magnetic sensor #3 at the top. This allows bench testing of the alarm box. For the final installation, you will remove the connections from the terminal block and use two-conductor wire to connect the sensors around the room being monitored.

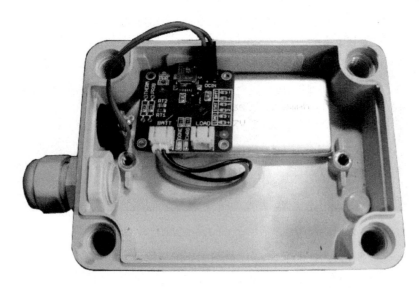

Figure 6-25. *Placing the battery and battery charging board in the alarm mounting box*

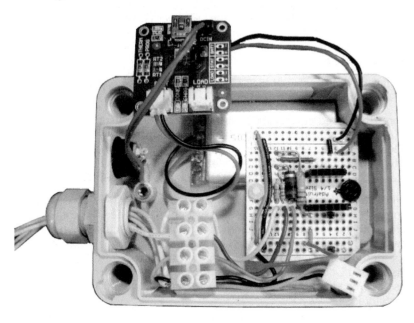

Figure 6-26. *Mounting the project in the box with wiring*

Code

The code for an alarm with three sensors on one branch is in Example 6-4 (also available from the repository for this book, directory *Chapter 6 Code*, subdirectory *Chapter6_03Alarm*).

This configuration has the sensors tied into resistors that are feeding Trinket pin #3 (which is also analog pin 3). You must program the Trinket out of the circuit, as pins #3 and #4 are shared with the USB connection.

The program uses the SoftwareSerial library to talk via the Bluefruit EZ-Link on Trinket pin #0 for transmit, pin #2 for receive. You can eliminate the requirement to specify a receive pin (and shrink the code slightly) with the SendOnlySoftwareSerial library, first used in Chapter 3. This would also allow you to use Trinket pin #2 (analog 1) for alarms, freeing pin #3, which is shared with USB. If you make such changes, be sure you make the corresponding changes with both the circuit and the code.

Trinket pin #1, which has the onboard red LED, is used as an additional annunciator to indicate which sensors are tripped. When the alarm is set to no alarms, the LED does not blink. It blinks from one to seven times, depending on which sensors are *tripped* (in the alarm state). The blink pattern is 1 for Sensor 1/PIR, 2 for Sensor 2, 3 for Sensor 3, 4 for 1 and 2, 5 for 1 and 3, 6 for 2 and 3, and 7 for all sensors tripped. If you decide to use Trinket pin #1 for other purposes, you can do so, but with the LED in-circuit you need to take that into account as previously discussed. For Trinket, the internal pull-up resistor, if enabled, is too weak. You may use a fairly low-value external pull-up resistor, nominally 1,000 ohms, if you decide to use the pin for sensors.

In the code for the project, enabling DEBUG (uncommenting the line `//
#define DEBUG` by deleting the `//` characters) will output the alarm values for the analog pin to the serial connection. You should do this once your circuit is together to ensure the values read by your circuit give values understood by the code as alarms are set off.

Trip each combination of sensors and record the value for the analog pin displayed on the serial line. Change the code line that has `uint16_t` values to the values you find. On the bench this process takes less than five minutes. If you have problems with the final install giving errors, try this process again because the resistance of the wiring may change the values.

Example 6-4. Sketch to operate the Trinket Alarm System

```
/* Trinket Alarm System */

#define SerialPin 0 // Serial via Bluefruit EZ-Link on this pin
#define LEDpin 1    // Use Trinket LED for displaying tripped sensors
```

```
#define SensorPin 3 // A3, which is pin #3, has resistor network to read
                     //   3 normally closed sensors
//#define DEBUG          ❶
#include <SoftwareSerial.h> // Software Serial library   ❷

SoftwareSerial Serial(2,0);    ❸
const uint8_t numSensors = 3; // number of sensors   ❹
const uint8_t states = 8;     // 2^numsensors
uint16_t values[8] = {541, 685, 661, 614, 840, 780, 776, 997};
char *textval[8] = {"Set","PIR", "2", "3","PIR2","PIR+3","2+3","All"};   ❺

void setup() {
  pinMode(LEDpin, OUTPUT);   // Set pin #1 to output for LED blinking
  pinMode(SensorPin, INPUT); // Set analog pin for input
  Serial.begin(9600);        // Send status information via serial
  Serial.println("Alarm System");   ❻
}

void loop()  {
  int8_t contact;         // Read alarm loops (returns -1 if a read error)

  contact = readContact(SensorPin);   ❼
  if(contact >= 1) {         // If any value greater than 0 (set),
    Blink(LEDpin, contact);  // we have an alarm! Blink LED corresponding to
    Serial.print("Alarm! "); // that sensor(s) and write to Bluetooth.
    Serial.println(textval[contact]);
  }
  else if(contact < 0) {    // A bad analog read was done. If you get errors
    Serial.print("Error");  // set DEBUG, walk test, record values, and
  }                         // update code with analogRead values.
  else {
    Serial.println("Set");  // Alarm is set (no sensors tripped)
  }
 delay(500); // We do not need to poll the sensors often (changeable)
}

int8_t readContact(uint8_t TrinketPin) {    ❽
// Returns the number corresponding to sensor values.
// TrinketPin is the analog pin on the Trinket (A1=#2, A2=#4, A3=#3).

  const int variance = 8; // Analog readings can vary, value for +-variance
  int contact = 0;
  uint16_t readval = 0;
  readval = analogRead(TrinketPin); // Check the pin

#ifdef DEBUG
 Serial.print(": Sensor read value: ");
 Serial.println(readval);
#endif

  for(uint8_t i=0; i<states; i++) { // If reading is near state value,
                                    //   return that state
    if(readval >= (values[i]-variance) && readval <= (values[i]+variance) )
      {
      return(i);
```

```
    }
  }
  return -1; // Value not one of the alarm system values
}

void Blink(uint8_t pin, uint8_t times) {    ❾
  for(uint8_t i=1; i<=times; i++) {
    digitalWrite(pin, HIGH);
    delay(85);
    digitalWrite(pin, LOW);
    delay(85);
  }
}
```

❶   If this line is uncommented, the debugging code will be included.

❷   Trinket pin #0 connects to the Bluefruit EZ-Link RX pin for serial
    communication. You may use a terminal program (such as the free-
    ware PuTTY for Windows) to get the alerts over Bluetooth. A
    Processing or Python script looking for alarms could also be used to
    automate monitoring.

❸   Serial transmission on Trinket pin #0, receive pin #2 (not used here).
    SendOnlySoftwareSerial may also be used for the transmission
    without receive. If you decide to use pin #2 for a sensor branch, this
    should be changed.

❹   This sets the number of normally closed sensors you'll be multiplex-
    ing on one analog pin. If you have two sensors, you can leave the one
    resistor open and adjust the text values.

❺   Text values are mapped to all the different states of sensor open and
    close to alert the monitor. This can also be done at the receive end if
    a number is passed instead of text.

❻   The receiver can look for this text to determine when the alarm has
    been turned on or restarted.

❼   If the alarm sensor configuration is changed (more sensors, etc.),
    change the readContacts function.

❽   readContacts should return 0 for no alarm, -1 for errors, or a number
    to indicate which alarm or alarms are going off. The logic can be
    changed here to add more sensors.

❾   This routine toggles a pin the number of times requested. This sketch
    uses it on an LED pin to indicate which sensors have been triggered.

If you uncomment the DEBUG line, remember to comment it out for your
final installation.

Final Assembly

Ensure you have made all the connections noted in Figure 6-23.

Program the Trinket out of the circuit, then place in the headers on the circuit board.

Check your power connections one more time. Plug the circuit board power connector into the LOAD connector on the LiPo charger. Lay the LiPo charger board on top of the battery.

Next, add the PIR sensor. Connect the PIR, ensuring the red wire is connected to the 5-volt power supply and the black wire connects to GND. The PIR has two orange variable resistors, one to adjust sensitivity and one for latch time. The latch time can be fairly short (we are sensing any change). The sensitivity should be turned down at first and may be adjusted when making the final installation.

You can place the PIR outside the enclosure but the nice clear case begs for an inside mounting scheme. There is enough clearance, but just enough. For best sensitivity, you may want to use a round hole for the lens to come out (with some sealant).

The Bluefruit is connected to the circuit board as follows:

- BAT+ (line into Trinket) goes to the Bluefruit Vin pin
- Ground (−) goes to the Bluefruit GND pin
- Trinket pin #0 goes to the Bluefruit RX pin

Tuck the Bluefruit transceiver in the case on its side. Figure 6-27 shows all the parts in the enclosure, a tight fit indeed.

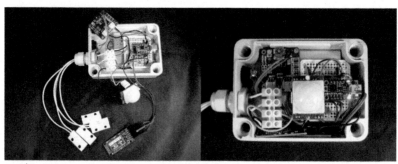

Figure 6-27. *Final assembly of parts into the project case shows a snug fit*

Test

Power up the circuit. The battery may charge with the CHRG light lit on the LiPo charger circuit board. The Trinket's green LED should be lit. If not, disconnect the power and check your connections.

When working correctly, the red LED on the Trinket will blink, showing the sensors that are tripped—this is normal, as the PIR sensor probably sees you moving and you might not have the magnets against the magnetic sensors. Place the magnets next to the sensors and aim the PIR away from you. The system should set up (no alarms are tripped). Then, when you move in front of a sensor, it should blink once a second or so, with the number of blinks indicating which sensors tripped, as follows:

No blinks
   Secure

One blink
   PIR tripped

Two blinks
   Magnetic Sensor 2 tripped

Three blinks
   Magnetic Sensor 3 tripped

Four blinks
   PIR and Magnetic Sensor 2 tripped

Five blinks
   PIR and Magnetic Sensor 3 tripped

Six blinks
   Both magnetic sensors tripped

Seven blinks
   All sensors tripped

If the PIR is not indicating correctly, move the orange potentiometers to reduce or increase sensitivity. If it will not "go off," reduce the latch time.

The system state is also broadcast via serial to the Adafruit Bluefruit EZ-Link at 9,600 baud. It transmits up to 10 meters (33 feet).

With a laptop or PC with a Bluetooth receiver, determine which serial port is connected to Bluetooth for your operating system. (For Windows, go to Control Panel→Devices and Printers; if you do not see Adafruit Bluefruit listed, use the Bluetooth program in the system tray (lower right, tiny blue B icon) to add it to your Bluetooth device list. You may have to press the pair button on the Bluefruit board to have it recognized.

See *http://bit.ly/Bluefruit_EZ-Link* for instructions on pairing to Bluefruit EZ-Link on Windows, Mac, or Linux.

To listen in on the serial stream, you should load a terminal program. For Windows, you may again use PuTTY; for Mac or Linux, use Terminal.

Open your terminal window and set it to work on a serial stream. Set your serial port (on Windows this is COMxx, where xx is the number of the port you found in Control Panel earlier; on Mac and Linux, it will be something like */dev/cu.AdafruitEZLink* followed by some other characters). The baud rate is 9,600. When you press the Open button, a black terminal screen will be displayed with white text output from the Bluetooth serial stream, similar to Figure 6-28.

Figure 6-28. *Serial terminal output from the Bluetooth sensor displaying alarm events*

The final project with the clear weatherproof case cover is shown in Figure 6-29.

Do a final test of the sensor function. Trip (activate) each sensor in turn and make sure you get the indications that were programmed in for annunciation. If you seem to have a magnetic sensor stuck on or that never activates, check your wiring.

You are now ready to perform the final installation. You will need to remove the magnetic sensor connections as these will be done with wires in the permanent location.

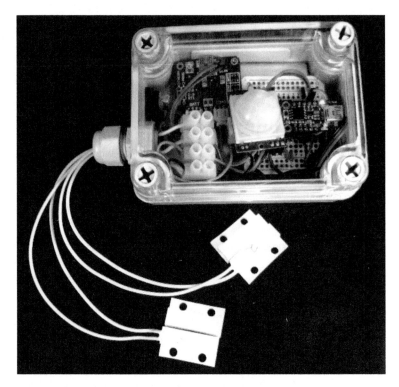

Figure 6-29. *The final alarm with cover, ready to install*

Troubleshooting

The Bluefruit should have both red and blue flashing LEDs when transmitting. If not, double-check that Trinket pin #0 is connected to Bluefruit RX and that you have GND and Vin connected to circuit ground and the BAT+ terminal, respectively.

For information on connecting to Bluefruit with different operating systems, refer to Adafruit's Bluefruit EZ-Link tutorial (*https://learn.adafruit.com/introducing-bluefruit-ez-link/pair-and-test*).

Going Further

Customizing this type of alarm is very straightforward, hardware-wise. You will need to ensure that the polling of all your sensors and branches is fully accounted for in software and change the annunciation type or text appropriately.

Alarm systems can be as complex as one can dream. You can create a disarm function, either wired or wireless, that says "I'm a good guy." You may consider a clock function that arms or disarms the system depending on the time of day. This could even be done by the monitoring PC, via a command received via Bluetooth. At a certain point, your expansion plans could exceed the capabilities of the Trinket. Still, this project demonstrates that quite complex alarm functions can be built within the Trinket's capabilities.

Bluetooth Communication

Bluetooth use continues to grow as communication with consumer electronics and peripherals expands. Bluetooth is one of the communication methods often used for the Internet of Things. The benefits of using Bluetooth instead of WiFi or other technologies include:

- The interface is usually via a simple serial connection.
- The range in some modes can be many meters.
- Coding is easy.
- Interfaces with a wide variety of commercial devices well.

Disadvantages include the following:

- It's relatively insecure.
- It's power hungry if power conservation is not built-in.
- It can be expensive for microcontroller use compared to simple radios.
- It's susceptible to radio interference in the 2.4 GHz band (WiFi, microwave ovens, etc.).

Adding Bluetooth radios to mobile products is standard now. The microcontroller industry is just catching up with consumer electronics. For years, Bluetooth hobby modules were expensive, were difficult to interface with, or were supplied in awkward form factors. This finally appears to be changing, due to new innovations. When shopping for a Bluetooth radio, be cognizant of your needs and the market before you decide on a solution.

The alarm project uses a modern board by Adafruit called the Bluefruit EZ-Link, shown in Figure 6-30. The "EZ" comes from the fact that the pinout and use are identical to the FTDI Friend board used in some other projects in this book. Adafruit also makes the Bluefruit in a standard Arduino shield format. Another standalone version, called Bluefruit EZ-Key, transmits keystrokes if pins are activated. The standalone board could actually make a very simple transmitter for switch closures, making for an alarm system where the sensor wires are replaced by Bluetooth.

Figure 6-30. *The Adafruit Bluefruit EZ-Link*

Most Arduino-compatible adapters are compatible with Bluetooth versions 2.1 and 3. Newer devices based on Bluetooth 4.0, including Bluetooth Low Energy (BT LE), are appearing on the hobby scene. BT LE provides features including better security, much lower power consumption, compatibility with newer consumer electronics, and more. Adafruit has introduced a Bluetooth LE module called Bluefruit LE based on the low-energy nRF8001 chip.

It is easier than ever to affordably add Bluetooth to your microcontroller projects, including those projects using a Trinket.

Trinket Toy Animal

Toy animals are a staple of children's toy boxes and Valentine's Day gifts. There is a resurgence of interest in making animals more interactive, as Makers desire items that move, blink and glow—maybe not quite like Teddy Ruxpin, but with features that will delight.

This project provides ideas for creating stuffed, paper craft, or other toy animals with characteristics you dream up. You can choose your animal, make it move, and have it make sounds. The project demonstrates how to use a servo motor, a piezo speaker, and a photocell to provide interactivity. Technically, you do not even have to use an animal—the techniques described here are suitable for simple robots or other items.

Choosing Your Animal

This project demonstrates animating a kiwi bird, shown in Figure 6-31. The electronics dictate the minimum size of the animal you choose. It may be as large as you want (with suitable changes for the servo and power). The

Beanie Baby bird here (named Beak) seemed well suited to animation. A moderate-sized Angry Bird toy may be a good alternative. You can use existing animals or fabricate your own with paper, cloth, or 3D printing.

Figure 6-31. *The animated toy*

Parts List

- Trinket 3V, Adafruit #1500 or Maker Shed #MKAD70
- Perma-proto quarter-sized breadboard (single), Adafruit #1608 or Maker Shed #MKAD48
- Cadmium sulfide photocell, Adafruit #161
- Piezo speaker/buzzer, Adafruit #160 or Maker Shed #MSPT01
- Micro servo, Adafruit #169, Maker Shed #MKMSERVO, or similar
- 3 x AAA battery holder, Adafruit #727 or Maker Shed #MKAD61
- Female header, Adafruit #598 or similar
- Rainbow female–female cable, Adafruit #793, #266, or similar

- Optional: Male header, Adafruit #392 or similar
- Hookup wire, Adafruit #289, #288, #290, or similar
- 10,000-ohm (10K) potentiometer (other values okay), Adafruit #562 or similar
- 1,000-ohm resistor
- Stiff house wire or stiff craft wire, several inches
- 3 AAA batteries
- Screw and nut
- Optional: Velcro

Tools

- Drill and bits
- Soldering iron and solder
- Diagonal pliers for cutting wires and header
- Wire strippers
- Scissors or seam ripper
- Glue
- Needle and thread

Circuit

The circuit connections, shown in Figure 6-32, are a bit simpler than in the other projects in this chapter. This project uses a Trinket 3V and will work with Gemma as well.

Start by soldering male header pins to the Trinket, as directed in "Preparing the Trinket" on page 17.

We'll use a quarter-sized Adafruit perma-proto board as the base. Hookup wire interconnects the components as shown in Figure 6-32. You'll use the 1,000-ohm (1K) resistor for the photocell circuit.

You can use female headers to make the Trinket removable from the protoboard, similar to other projects in this chapter. Cut two five-pin sections. You can use male headers to connect off-board components: two two-pin and one three-pin section connect the servo, piezo, and photocell. Make the connections via strips of female–female jumper wires, or you can use any flexible wire. Using headers allows for easy removal of the components from the project as the build progresses or while programming the Trinket.

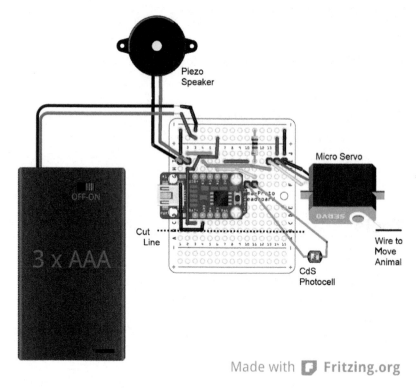

Figure 6-32. *The Trinket Toy Animal wiring diagram*

The perma-proto circuit board may be cut at row B to make the board smaller if your animal is more petite than can be accommodated by the full board.

The servo is mounted onto the perma-proto board to provide a base for moving the servo (it needs something to push against). You can mount it in many ways, but it generally needs a firm mount to achieve the movement you may want. I enlarged the proto mounting hole slightly with a drill bit to allow a small screw to pass through at the correct place.

Circuit Variations

No single build is perfectly suited to every project. You may consider the following changes to the project:

Trinket
You can use a Trinket 5V with a 5-volt supply, perhaps 4 AAA or AA batteries. The servo is generally 5-volt, so be careful running at 6 volts (check the servo datasheet). You can put a silicon diode in series with a 6-volt supply to drop the voltage by about 0.7 volts. A Gemma would

also work well, especially if you want to use a LiPo battery, but proto mounting would be more of a challenge.

Photocell

If you want the circuit to always run when switched on, you can eliminate the photocell and 1,000-ohm resistor. You will need to remove the `analogRead` function call and the `if` statement in the code that acts on the reading.

Speaker

If you want louder sounds, you can feed the Trinket pin #1 sounds to a transistor or amplifier to drive a more conventional speaker.

Servo

If you use a larger servo, you may need more battery capacity. If you do not need your animal to move, you can eliminate the servo and the wire from Trinket pin #0, power, and ground. You could then trim the amount of proto board required by cutting at line 12.

Battery

The Trinket 3V runs well on LiPo batteries stocked by Adafruit. The 3.7V 1,200 mAh is a good size and capacity, but smaller sizes work as well (with reduced run time). If you use a LiPo, you may want some type of on/off switch. AA and AAA battery packs may have the switch built in. I fashioned an ad hoc JST male connector for a LiPo with two spare header pins in the upper-left corner of the circuit board. A Gemma or Trinket version 1.1 with a soldered JST connector would already have the JST battery connection, if you want to consider that as an option.

Code

The software in Example 6-5 is designed for a bird. You can change the chirp function for one of the other sounds discussed in Appendix A.

The code reads the photocell, and if the value is less than a certain amount (e.g., if you pet the animal, lowering the light level), the animal makes sounds and moves. You can change the sensitivity of the photocell, currently 800, to a value between 200 and 1100, depending on the ambient light expected.

The register/interrupt code allows the servo to be refreshed periodically to have it stay where it is commanded to. The standard Arduino IDE Servo library does not work for Trinket/Gemma, so the Adafruit SoftServo library is used. See "ATtiny-Optimized Libraries" on page 39 for library locations and "Installing Libraries" on page 41 for instructions on installing libraries.

The code may be downloaded from the GitHub repository for this book (directory *Chapter 6 Code*, subdirectory *Chapter6_05Animal*).

Example 6-5. The Trinket Toy Animal main sketch

```
/*  Trinket Toy Animal */

#include <Adafruit_SoftServo.h> // SoftwareServo (works on non-PWM pins)

#define SERVO1PIN 0     // Servo control line (orange) on Trinket pin #0
#define SPEAKER   1     // Piezo speaker on pin #1
#define PHOTOCELL 1     // CdS photocell on pin #2 (analog pin 1)

Adafruit_SoftServo myServo1; // Create servo object
int16_t servoPosition;       // Servo position

void setup() {
  OCR0A = 0xAF;            ❶
  TIMSK |= _BV(OCIE0A);

  servoPosition = 90;            // Tell servo to go to midway
  myServo1.attach(SERVO1PIN);    // Attach the servo to pin #0 on Trinket
  myServo1.write(servoPosition); // and move servo
  delay(15);                     // Wait 15 ms for servo to reach the position
  pinMode(SPEAKER,OUTPUT);    ❷
}

void loop() {
  uint16_t light_reading;
  if(servoPosition != 0)    ❸
    servoPosition = 0;       // If it's up, go down & vice versa
  else
    servoPosition = 180;
  light_reading = analogRead(PHOTOCELL);    ❹
  if(light_reading < 800) { // If the photocell is dark, we're petting
    chirp();                //    the animal, so make sound ...
    myServo1.write(servoPosition); // and tell servo to move
  }
  delay(1000); // Wait between chirps (can be changed)
}

void chirp() {  // Generate the Bird Chirp sound
  for(uint8_t i=200; i>180; i--)
  playTone(i,9);
}

void playTone(int16_t tonevalue, int duration) {    ❺
  for (long i = 0; i < duration * 1000L; i += tonevalue * 2) {
    digitalWrite(SPEAKER, HIGH);
    delayMicroseconds(tonevalue);
    digitalWrite(SPEAKER, LOW);
    delayMicroseconds(tonevalue);
  }
}

volatile uint8_t counter = 0;    ❻
SIGNAL(TIMER0_COMPA_vect) {
  // This gets called every 2 milliseconds
```

```
  counter += 2;
  // Every 20 milliseconds, refresh the servo!
  if (counter >= 20) {
    counter = 0;
    myServo1.refresh();
  }
}
```

❶   This code sets up the servo refresh timer.

❷   If you forget to set the speaker as an output, you won't hear a sound!

❸   The servo position is changed every loop, but it's only activated if you're petting the animal.

❹   Read the photocell. If the light level is below a certain value, something is blocking it; this is how we determine the animal is being petted.

❺   This toggles a pin to play a tone on a piezo speaker.

❻   This routine refreshes the servo every 20 milliseconds as required.

Preparing the Toy

The animal I chose is a classic Beanie Baby toy. The code makes a chirping sound from the piezo, and the servo moves a wire that animates the beak.

As shown in Figure 6-33, the seams on the bottom of the bird must be carefully cut with a seam ripper or scissors. Remove the plastic "beans" and fluff stuffing, saving the fluff for later. Cut a piece of stiff wire, which you'll stuff up into the head and into the beak shortly. A spare piece of house wiring (10 to 12 gauge) works well, but anything fairly stiff but bendable will do.

Push the photocell leads through the seam at the forehead, as shown in Figure 6-34. Use a dot of glue to attach the photocell to the animal fur. Now for the tricky part: carefully solder a piece of two-conductor flexible wire (like rainbow wire with connectors cut off one end) to trimmed photocell leads. You do not want to burn the fur with your soldering iron. Protect the soldered ends from flexing or touching with hot glue, sugru, or other material.

Drill a hole in the servo horn to match the diameter of the stiff wire for the head. The single-arm horn works fine, but the double-arm or circular horn will work as well. Strip and bend the end of the wire so it will fit on the servo horn and stay somewhat secure, but still move when needed.

Push the wire up through the beak. Bend the other (stripped) end slightly and thread it into the servo horn.

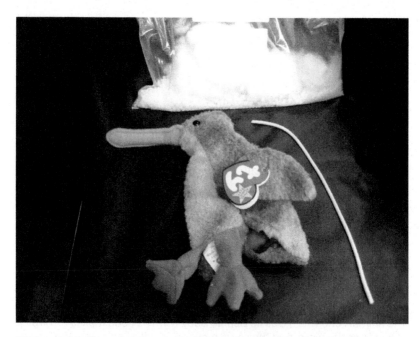

Figure 6-33. *Preparing the stuffed animal by removing the stuffing (carefully)*

If you want to maximize the speaker sound, you can place it so the piezo hole is exposed to the exterior of the animal. The fabric does not mute the chirp sound by much with such a shrill tone.

Make all the connections and test. If something does not work, remove the circuit and check the connections and the battery. Remember the photocell must be covered for the sound and movement to start; light will make it stop.

Take some of the interior fluff and stuff the head. Fit the electronics and wire to the servo together and place them inside the bird. You can use some additional stuffing to make the bird fatter, but do not impede the wire from moving. The "beans" should not be reused in the project as they may leak all over the place.

Once you have everything assembled to your satisfaction, you can stitch the body. However, you need to be able to switch the batteries on and off and change them when depleted. Sewing on some Velcro is a good way to seal things.

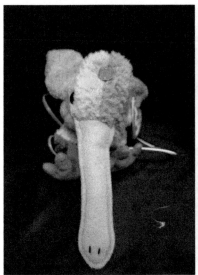

Figure 6-34. *Placing the photocell and servo wire inside the animal*

Use

When switched on in the light, the bird should sit still. Place your hand so that it blocks the light (petting the head), and the head should move and you should hear the chirp sound.

The fun part of this project is that you can use these techniques to animate any animal or object any way you want. Do you want an animated cat, dog, or owl? It is possible. Paper, plastic, 3D-printed objects—they all work very well.

You can also animate other objects, like small robots. For a more robust robot, the next project uses a sensor and two servos to make a robotic rover.

Trinket Rover Robot

The science, technology, engineering, (art,) and math (STEM/STEAM) curriculum is gaining momentum in education. It is exciting to see so many new Makers and engineers learning how fun it is to Make! Rick Winscot wanted to design a low-cost robot that anyone with access to a 3D printer could build.

Here is his solution: an autonomous micro rover based on the Trinket, shown in Figure 6-35.

Figure 6-35. *The Trinket Rover Robot*

When Rick started designing the rover, he ran into a significant obstacle trying to find inexpensive tracks or treads. He had some success 3D printing them with flexible filament, but the total cost was too high. That is when he stumbled onto *chain bracelets* (Figure 6-36), from Oriental Trading. You can buy a dozen for less than $10, which will supply treads for six rovers.

Figure 6-36. *Chain bracelets repurposed as rover treads*

The pattern is slightly different on the inside and outside, as you can see in the right-hand image in Figure 6-36: rounded (left) and flat (right). The rounded side fits in the wheel/sprocket perfectly.

Parts List

- Trinket 3V, Adafruit #1500 or Maker Shed #MKAD70
- Tiny breadboard, Adafruit #65 or Maker Shed #MKKN1-B
- Maxbotix ultrasonic rangefinder LV-EZ1, Adafruit #172

  OR

  Parallax PING))) ultrasonic sensor, Parallax #8015 or Maker Shed #MKPX5 (this is the sensor used in the pictures; using others may require additional work)

  OR

  Grove ultrasonic rangefinder, Speed Studio #SEN1073P or Maker Shed #MKSEEEE27
- 4xAA battery holder, Adafruit #830
- Two continuous rotation (CR) micro-sized servos. You can convert two standard micro servos per *http://bit.ly/continuous_rotation* (advanced) or buy them from a robotics supplier such as Robot-Shop.com (#RB-Fit-02). Unfortunately, Adafruit's CR servos are standard size, not micro
- Rainbow female–male jumpers, 6" (150 mm), Adafruit #826
- Extra-long male–male pins, Adafruit #400 or similar
- 3 M3 10 mm screws
- 1 M3 hex nut
- Double-sided foam tape
- Chain bracelets, Oriental Trading #IN-13605773 or similar (color availability may vary by season)
- Plastic cement or 5-minute epoxy

Tools

- 3D printer (or send files to a 3D print shop)
- Drill and bits
- Screwdriver
- Sandpaper
- Diagonal cutters
- Optional: M3 thread tap

3D Printing

The rover is composed of seven plastic parts. (Rick says the moustache is *not* optional.) The model archive includes a small and large chassis and three different sonar mounts. Depending on the printer, you may be able to squeeze all of the parts into a single print.

You can download the 3D (*.stl*) files at *http://bit.ly/rover_3D_files*.

3D printers are now located in many places, including schools and Maker-spaces. A growing number of companies also are providing 3D printing services.

When you get your parts, inspect them for rough places. Use diagonal cutters if any stray filaments protrude on a piece. Use sandpaper on any piece that appears rough. A good print will not have much variation in the layers. If you print your own and find the pieces uneven, you may wish to calibrate the printer or check your material.

Build

You might need to tap the axle holes. Alternatively, you could warm/soften the plastic with heat and thread the hole with the screw, but be careful! Make sure that the wheel on the front does not bind (see Figure 6-37). You might need to sand around the axle a bit. Screw the wheels on per Figure 6-38.

The mounting hole in the wheel is purposefully small to allow for the widest possible screw size. Widen the hole with a drill bit that is slightly larger than the threads of your screw.

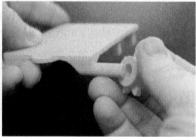

Figure 6-37. *Fitting the wheels onto the rover*

Insert an M3 hex nut and screw the sonar mount to the chassis, as shown in Figure 6-39.

You'll need pilot holes (Figure 6-40) to mount the servo to the chassis. A 1/16-inch bit will do nicely.

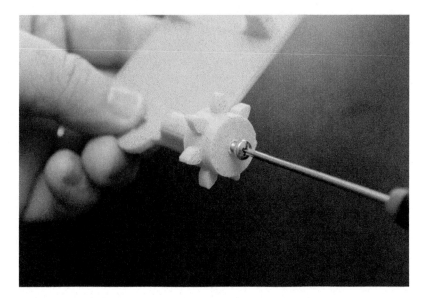

Figure 6-38. *The screw holds the wheel on the body*

Go slowly, securing the servos to the chassis with screws, as shown in Figure 6-41. If you hear any kind of cracking, you might want to carefully warm or soften the plastic a bit before proceeding.

Now to work on the rear wheels. Grab a dual-arm servo horn and trim it to fit a wheel. You can use model cement to attach the servo horn to the wheel (Figure 6-42), but it will take an hour or so before it's dry enough for final assembly (Figure 6-43); 5-minute epoxy might be a better alternative.

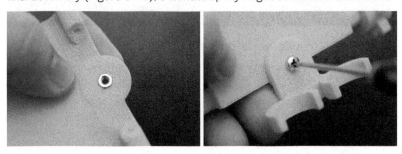

Figure 6-39. *Installing the sensor mount*

Repeat these steps for the other wheel. When the wheels are fully dry, mount the wheels to the servos with the screws that came with the servos. Carefully place the treads onto the wheels.

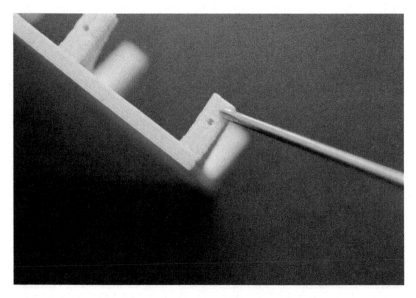

Figure 6-40. *Drill the pilot holes to mount the servos on the chassis*

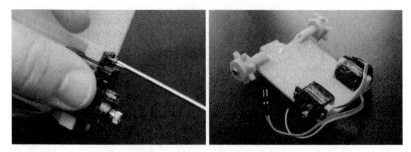

Figure 6-41. *Screwing the servos onto the rover chassis*

Figure 6-42. *Gluing servo horns onto the powered wheels*

Figure 6-43. *The finished powered wheel*

Wiring

All three distance sensor options in the parts list work fantastically. All you need is GND, V_{CC}, and one pin for the measurement signal. Snap the appropriate sonar into the mount, as shown in Figure 6-44.

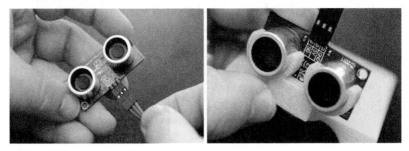

Figure 6-44. *Connecting and mounting the sensor*

Insert extra-long male–male pins, shown in Figure 6-45, into the female servo cable to convert to male pins.

The circuit diagram is in Figure 6-46. Wire all the required connections. With the servos, remember that ground is often brown and signal is orange. Red is power.

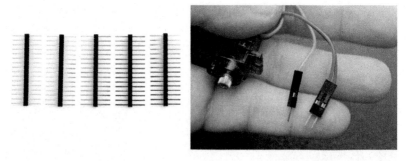

Figure 6-45. *Make the servo cable pins male for breadboard use*

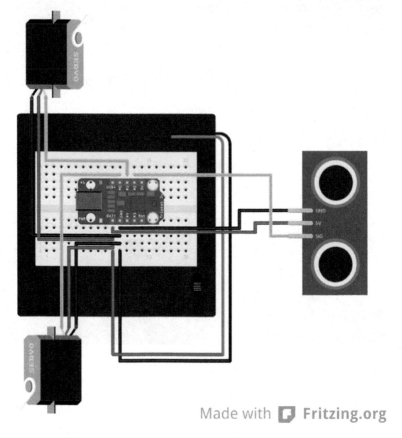

Made with ⬛ Fritzing.org

Figure 6-46. *Wiring diagram for the Trinket Rover*

Place batteries in the holder. Secure the battery holder to the breadboard with double-sided tape.

Code

You'll use the Adafruit SoftServo library for this project. See "ATtiny-Optimized Libraries" on page 39 for library locations and "Installing Libraries" on page 41 for instructions on installing libraries.

The sketch in Example 6-6 uses the Parallax Ping))) distance sensor and is a starting point. You can change it to use a different sensor or to add more behaviors.

You can download the code from the GitHub repository for this book (directory *Chapter 6 Code*, subdirectory *Chapter6_06Rover*).

Example 6-6. Sketch to operate the Trinket Rover Robot

```
/* Trinket Rover    Version 1.0 Rick Winscot */

#include <Adafruit_SoftServo.h>

#define SERVO1PIN 0 // Servo control line (orange) on Trinket pin #0
#define SERVO2PIN 1 // Servo control line (orange) on Trinket pin #1
Adafruit_SoftServo servo_left, servo_rght;

const int sonar = 2;        // Sensor on Trinket pin #2
const int left_speed = 75;    ❶
const int rght_speed = 90;
const int obstacle = 8;       ❷
const int back_track = 100;   ❸
long duration, inches, cm;

void setup() {
  servo_left.attach(SERVO1PIN); // Attach servos...
  servo_rght.attach(SERVO2PIN); //   and off we go!
}

void loop() {
  servo_left.write(left_speed - cm); // Setting servos
  servo_left.refresh();              // in forward motion
  servo_rght.write(rght_speed + cm);
  servo_rght.refresh();
  delay(15);
  duration = 0;    ❹
  inches  = 0;
  cm      = 0;
  pinMode(sonar, OUTPUT);   ❺
  digitalWrite(sonar, LOW);
  delayMicroseconds(2);
  digitalWrite(sonar, HIGH);
  delayMicroseconds(5);
  digitalWrite(sonar, LOW);
  pinMode(sonar, INPUT);    ❻
  duration = pulseIn(sonar, HIGH);
```

```
inches = microsecondsToInches(duration); // Convert time into distance
cm = microsecondsToCentimeters(duration);
if ( cm < obstacle ) {     ❼
  for (int i = 0; i < back_track; i++) {    ❽
    servo_left.write(150);
    servo_left.refresh();
    servo_rght.write(50);
    servo_rght.refresh();
    delay(15);
  }
}
}

long microsecondsToInches(long microseconds) {
  return microseconds / 74 / 2;    ❾
}

long microsecondsToCentimeters(long microseconds) {
  return microseconds / 29 / 2;    ❿
}
```

❶   This is a moderate forward speed for both servos. Given the servos' orientation, one will be going forward, and the other backward. You may need to adjust these slightly to get the rover to move straight forward.

❷   This is the number in centimeters that the rover will reverse and try to navigate around.

❸   Multiplier used to determine how far the rover will back up.

❹   Establish variables for the duration of the ping, and the distance result in inches and centimeters.

❺   The ping is triggered by a HIGH pulse of 2 or more µseconds. Give a short LOW pulse beforehand to ensure a clean HIGH pulse.

❻   The sonar pin is used to read the signal from the PING))): a HIGH pulse whose duration is the time (in microseconds) from the sending of the ping to the reception of its echo off an object.

❼   back_track * delay(15) is the distance the rover will back up during obstacle avoidance.

❽   These are preselected numbers for moving the rover in a defined way. The rover always performs the same maneuver in avoiding an obstacle. Replace the code here with your own to define alternative obstacle avoidance behavior.

❾   According to the Parallax datasheet (http://bit.ly/Parallax_data sheet) for the PING))), there are 73.746 microseconds per inch (i.e., sound travels at 1,130 feet per second). This gives the distance traveled by the ping, outbound and return, so divide by 2 to get the distance to the obstacle.

⓾ The speed of sound is 340 m/s, or 29 μs/cm. The ping travels out and back, so to find the distance to the object, take half of the distance traveled.

Program the Trinket and plug it into the breadboard. Be sure the Trinket is oriented correctly and in the correct pin locations.

Switch on the rover. The Trinket should power on, and it should move and avoid obstacles! If you have problems, check your connections and programming.

Going Further

Other creative Makers have built robotics projects using the Trinket. Some use it as the main controller; others as a servo, a sensor processor, or both. An obvious use would be as an add-on processor for the Raspberry Pi. The Pi cannot do real-time processing and driving as well as the Arduino because of overhead in the Linux operating system. The addition of the Trinket can be a perfect fit. You can interface a Pi and Trinket via Software-Serial on Trinket pins #3 and #4; just be sure to program the Trinket without those pins being connected.

The next project highlights another facet of the Trinket: its ability to produce better sounds than those in previous projects. To provide that functionality, we'll use the final capability of the Universal Serial Interface: Serial Peripheral Interconnect (SPI).

SPI Communications

We've explored communications with the Trinket throughout this book. In Chapter 5, we looked at serial communication and I^2C. The final mode of serial communications offered by the ATtiny85's USI interface is the Serial Peripheral Interface. We'll use SPI in the next project to move data at high speeds between the Trinket and a flash memory chip.

In SPI, devices communicate in a master/slave mode where the master device initiates the data transfer. Multiple slave devices are allowed with individual slave select lines, as shown in Figure 6-47. Each device uses at least four digital pins for the connection:

SCLK
    Serial Clock (output from master)

MOSI
    Master Output, Slave Input (output from master)

MISO
    Master Input, Slave Output (output from slave)

SS
> Slave Select (active low, output from master)

You might see the connections labeled differently on some devices:

SCLK
> SCK, CLK

MOSI
> SIMO, SDO, DO, DOUT, SO, MTSR

MISO
> SOMI, SDI, DI, DIN, SI, MRST

SS
> nCS, CS, CSB, CSN, nSS, STE, SYNC

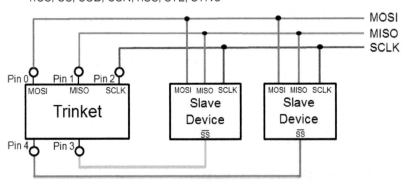

Figure 6-47. *SPI bus connections to a Trinket*

As there is no standard governing SPI, you may find that instead of connecting MOSI master to MOSI slave, the device is looking for SDO/DO/DOUT master connecting to SDI/DI/DIN on the slave, which may seem backward. This is a naming convention mismatch rather than an incorrect diagram. If you find an SPI circuit that is not communicating, check the wiring and datasheets: you may have to switch the two lines to get communications working.

The select pin is used in lieu of a slave address, as in I2C. The benefit is that the communication is very fast and not limited to 8-bit words. Disadvantages include use of more pins to implement SPI, that only one master may be on the bus, and that it is limited to short distances. Because it is not defined as a formal standard, you must refer to datasheets of specific devices on timing and use.

Not all circuits require every connection. An SPI display may only require data in and not return data out (or it may be optional). If you're dealing with a single device connection, this kind of arrangement can save you a pin you can use for other purposes.

SPI is implemented in a wide range of devices. Many of the Adafruit display backpacks we used in I²C mode also work in an SPI mode.

If you believe you are having issues implementing an SPI interface, there are several logic tools that can decode SPI bus signals. The popular Bus Pirate device by DangerousPrototypes (*http://dangerousproto types.com*) (Adafruit #237) is among these.

More discussion on use of SPI on the ATtiny85 is at *http://bit.ly/ SPI_on_ATtiny85*. The Arduino reference for SPI (*http://arduino.cc/en/ Reference/SPI*) is more for larger Arduinos. I also recommend Wikipedia (*http://en.wikipedia.org/wiki/Serial_Peripheral_Interface_Bus*) for more information.

Using SPI on a Trinket is not difficult, but not common. Using four pins, it does not leave many pins for other functions. The next project, the Trinket Audio Player, will demonstrate using SPI on the Trinket.

Trinket Audio Player

Our final project is a very nice design, again by Phillip Burgess. The Trinket might be thought of as a tiny subset of a "real" Arduino: less RAM, less code space, fewer input/output capabilities. But the Trinket has a couple of tricks up its sleeve: capabilities its larger brethren do not have. One of these is a high-speed PWM mode. With just a few extra components, this mode can be used for audio output—not simply piezo beeps and buzzes, but actual sampled digital sound!

You could make an electronic greeting card with your own customized message or song, add a background soundtrack to a model train diorama, or create the world's smartest whoopee cushion.

The circuit in Figure 6-48 will play audio from an SPI flash chip. It cannot play MP3 files, but you can convert sounds in MP3 format that fit onto the chip (up to about 65 seconds of playback).

Why this weird flash memory chip and not an SD card? Good question! There are a couple of reasons:

- The flash chip is super affordable, so you can make it a permanent part of a small project.

- The flash chip is in the form of an easy-to-use dual inline package (DIP). If you used an SD card, you would need to buy a special break-out board.

- Reading a FAT-formatted SD card with the Trinket's tiny microcontroller is incredibly difficult; a single SD block fills the Trinket's entire RAM. There are projects that do this, so it is not impossible, but it is nonetheless challenging.

Figure 6-48. *The Trinket Audio Player*

Parts List

There are two phases to this project. In the first, you will load sound data onto a flash memory chip using a regular Arduino. In the second you'll use a Trinket to play it back. Both stages have some parts in common:

- Winbond 25Q80BV serial flash memory (1 MB), Adafruit #1564— according to the datasheet (*http://www.adafruit.com/datasheets/ W25Q80BV.pdf*), this can store about one minute of music or two minutes of voice, depending on quality
- Half (or full) breadboard, Adafruit #64, Maker Shed #MKKN2, or similar
- Breadboard jumper wires, Adafruit #153 or Maker Shed #MKSEEED3, or any solid-core wire

For the loading stage:

- Arduino Uno, Adafruit #50 or Maker Shed #MKSP99, or similar board
- 0.1 microfarad (µF) capacitor, Adafruit #753 or similar
- 3 470-ohm resistors, RadioShack #271-1317 or similar
- 3 1,000-ohm resistors, RadioShack #271-1321 or similar

- Optional: LED (any color, Adafruit #299 or similar) and 220-ohm resistor (RadioShack #271-1313) for a status indicator

Not all of these parts are available from one vendor. You may be able to swap some out for different parts you already have on hand or can acquire locally; "Loading Sounds" on page 188 has some guidance on alternative parts selection.

For the playback stage:

- Trinket 3V, Adafruit #1500 or Maker Shed #MKAD70
- Female headers, Adafruit #598
- Small "mint-tin" perma-proto board, Adafruit #1214
- 10,000-ohm (10K) potentiometer, Adafruit #562 (large) or #356 (small)
- Hookup wire, Adafruit #289, #288, #290, or similar (such as Maker Shed #MKEE3)
- 8-pin DIP chip socket (RadioShack #276-1995 or similar)
- 2 0.1 μF capacitors (Adafruit #753 or similar)
- 1 10 μF capacitor (RadioShack #272-1025 or similar)
- 1 68-ohm resistor (RadioShack #271-1106 or similar)
- 3xAAA battery holder, Adafruit #727 or Maker Shed #MKAD61 (or you can use USB power)
- 3 AAA batteries
- Headphone jack, Adafruit #1699, and headphones or portable amplified speakers, Adafruit #1363 or similar

  or

  Audio amplifier board, Adafruit #1552, and 4-ohm (or 8-ohm) speaker, Adafruit #1314 or similar

Here, too, there is a lot of wiggle room for parts: not everything needs to be a precise value. "Sound Playback" on page 193 has some guidance on alternative parts selection.

Tools

- Soldering iron and solder
- Diagonal cutters to trim component leads
- Wire strippers

- Very optional: Oscilloscope to see data waveforms on the pins

Software

You will need sound files in WAV format. You can search for downloadable examples on the Internet (movie quotes, cartoon sounds, etc.). You can also record or convert something from your music collection using software such as Audacity (available as a free download (*http://audacity.sour ceforge.net/*)).

This project uses both the Processing language and the Arduino IDE. They look similar when running, which can lead to confusion. Make sure you are loading the right code in the right IDE! Processing is for writing code to run on your computer, while Arduino is for writing microcontroller code.

Download version 2.0 (or later) of Processing from *http://processing.org*— the software in this section will not work with version 1.5, if you currently have that installed.

The only library required for this project is the TinyFlash library. You can download it from *https://github.com/adafruit/Adafruit_TinyFlash/*. Download and install it as you've done for other libraries, per Chapter 4. When you've properly installed this library, you should have access to the code in the Arduino IDE via the menu item File→Sketchbook→Libraries→Adafruit_TinyFlash.

The *examples* folder included with the library contains all the code for this project; there is nothing else to download.

Loading Sounds

Sound files for this project need to be in WAV format, uncompressed (PCM), 8- or 16-bit resolution. Mono, stereo, or multi-channel are all acceptable. The software used here will automatically convert to 8-bit mono if needed.

If your audio is in a different format, you can convert it with a tool such as Audacity (*http://audacity.sourceforge.net/*) or Adobe Audition (*https:// creative.adobe.com/products/audition*), if you don't already have a conversion utility on your computer. (Even iTunes can convert to WAV, if you tweak the import settings—see Figure 6-49.)

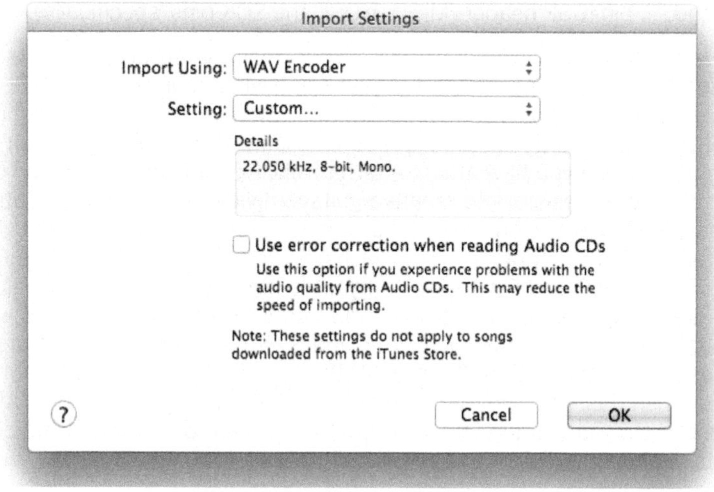

Figure 6-49. *Encoding audio for the music player*

For voice recordings, 8 KHz is often a sufficient sample rate. For music, use 16 KHz or more. Generally, higher sampling rates will produce better-sounding audio, but it requires more space. Also, the way the playback circuit works, there are diminishing returns above 25 KHz. You can experiment with the settings.

The Winbond flash chip you're using has a capacity of 1,048,576 bytes (1 megabyte, often called "8 megabit"). You'll use 6 bytes to store data about the length and sampling rate of the audio, leaving 1,048,570 bytes for the audio data itself. Each byte is one audio sample.

To estimate the maximum duration of audio you can store on the chip:

    Max. duration (in seconds) = 1,048,570 ÷ sampling rate

So, with 16,000 Hz (16 KHz) music:

    1,048,570 ÷ 16,000 = ~65.5 seconds

If your source audio file is too big for the available space, the sound will be truncated to fit when you load it on the chip.

Chip Loading Circuit

Because the Trinket does not support traditional serial I/O, you'll use a regular Arduino board to transfer data to the Winbond memory chip. Later, the memory chip will be moved to a playback circuit.

The Arduino Uno is suggested because it has easy access to the SPI pins used to communicate with the flash chip. This can be done with a

Leonardo, Mega, or other board, but you will need to adapt the wiring to use the six-pin ICSP header (MISO, MOSI, and SCK pins, specifically).

The serial flash chip is a 3.3-volt device, whereas the Arduino is 5 volts. To avoid unleashing the Blue Smoke Monster (frying your chip), it is necessary to power the chip from the Arduino's 3.3V (not 5V) pin, and then use level-shifting circuitry to drive the control lines. There are chips that do this; use them if you have them. Alternatively, a simple approximation can be made using resistors to create a voltage divider: connect each output signal pin from the Arduino to a 470-ohm resistor, with the other end connected both to an input pin on the flash chip and to a 1K resistor to ground (see Figure 6-50 and Figure 6-51). If you do not have exactly these values of resistors on hand, that is okay. You can substitute other values with approximately a 1:2 ratio, such as 1K and 2.2K (or use two 1K resistors in series for the latter). 1K/2.2K is about the upper limit on values; do not go higher than this. It will not harm anything; it just will not work reliably. Also, note that the data output line back into the Arduino (pin 12, blue wire) connects directly; there are no resistors on this line.

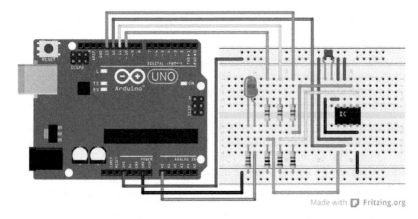

Made with **Fritzing.org**

Figure 6-50. *Wiring diagram for the audio loading circuit*

Add an LED and a 220-ohm resistor between Uno pin A0 and GND to provide a simple status display. The LED will blink to indicate an error, and it flickers during the data transfer. This is not essential and can be left out of the circuit if you do not have the parts.

Finally, you'll put a 0.1 μF capacitor between the flash chip's V_{CC} supply and GND. In a pinch you can get by without this, but it is good form to have it there; it keeps the electrical gremlins away.

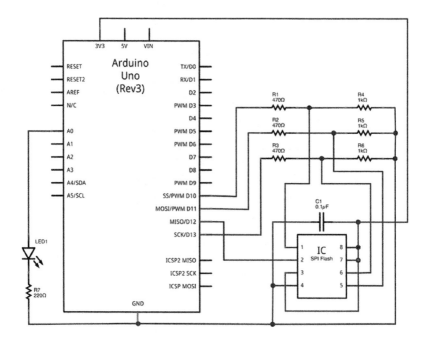

Figure 6-51. *Schematic for the audio loading circuit*

Breadboarding works fine for occasional use. Knowing that a whole lot of chips would be programmed while debugging this project, Phil wired all of the components on an Arduino proto-shield (Figure 6-52), with a socket, so he could easily swap out the flash chip.

Now launch the Arduino IDE and load the AudioLoader sketch: File→Sketchbook→Libraries→Adafruit_TinyFlash→AudioLoader.

Select the board type and serial port from the Tools menu, and upload this code to the board. If you do not have the flash memory chip installed (or if it is positioned incorrectly, or turned around), the status LED should blink. If it is working properly, you will not see anything from the LED, but you can check the serial monitor (at 57,600 baud). It should display:

```
HELLO
1048576
```

If not, something may be amiss with your wiring. Double-check all the connections against the schematic.

Figure 6-52. *The audio load circuit on an Arduino proto-shield*

Transferring Audio

Close the Arduino serial monitor if you still have it open; the other code will not work if it is there.

Launch Processing and load the AudioXfer sketch. It is inside the *Adafruit_TinyFlash* folder that you downloaded earlier, in a subfolder called *Processing*. (Sorry to make you hunt through the Arduino folder for this, but it was less troublesome than requiring a separate download!)

 Processing and the Arduino IDE look very similar. If you encounter strange errors, make sure you are loading the right code in the right environment.

When you run the AudioXfer sketch, all serial ports on the system are scanned until an Arduino running the AudioLoader sketch is identified. If you already know the name of the port (previously selected from the Tools→Serial Port menu in the Arduino IDE), there is a line you can uncomment to open this port directly and bypass the whole lengthy port scan:

```
portname = "/dev/tty.usbmodem1a1331"; // bypass scan
```

This is just an example port name you might see on a Mac or Linux. In Windows, it might be something like "COM6".

If the software detects an Arduino running the AudioLoader sketch, and if it reports a flash memory chip is connected, you will be prompted to select a WAV file to transfer. When you select a WAV file and click Open, the chip is erased. There is no undo.

AudioXfer is a pretty bare-bones program; other than file selection, there is no fancy user interface. It just prints text to the console. You will see a long line of dots (and the LED will flicker) as data is transferred to the Arduino and written to the chip. It can take several minutes to load a 1 MB chip to capacity, so you may want to test with just a short sound at first.

If all goes well, the software reports **done!**. If an error was encountered, you will instead see a message with some indication of the problem.

Once a sound is successfully loaded, disconnect the Arduino from the USB cable, remove the flash memory chip from the breadboard or socket, and then move it over to the playback circuit, which you'll build next.

Sound Playback

The ATtiny85 chip at the heart of the Trinket has the novel ability to produce a 250 KHz 8-bit PWM signal. That's four times what the Arduino Uno can muster. You'll build a low-pass filter circuit to smooth that "square" PWM into a usable audio waveform, shown in Figure 6-53.

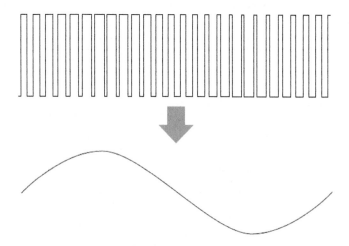

Figure 6-53. *Smoothing a PWM signal into an audio signal*

You can make a basic low-pass filter from just a capacitor and resistor. You need to know the filter's cutoff frequency—frequencies below this pass through, while higher frequencies (like the PWM signal) are *attenuated*. A rule of thumb with PWM audio is that the highest usable audio frequency (the cutoff frequency) is about one-tenth the PWM rate. We've established that this is 250 KHz, so a good cutoff would be 25 KHz.

There is a relationship between the capacitor and resistor values and the resulting cutoff frequency. Given any two values, the third may be computed. Already having many 0.1 microfarad (abbreviated µfarad or µF) capacitors around, Phil just needed to know the corresponding resistor value required to achieve the desired 25 KHz cutoff (you could also do it the other way: start with some resistor you have around, then determine a suitable capacitor). Rather than getting bogged down with the math, you can just whip out the Adafruit Circuit Playground app for iOS (select "Circuit Calculators," then "RC Cutoff Filter"). You can also search in Google for "low-pass filter calculator" resources on the Web and plug in the two known values you have.

For a 25 KHz cutoff and 0.1 µfarad capacitor, the Circuit Playground app suggests a 63-ohm resistor (Figure 6-54). That is not a standard value you'll find anywhere, so select the next common size up from there: 68 ohms. If you have a 75-ohm resistor around, that is close enough; this is not precision work.

The filtered output is then fed into a 10K potentiometer for volume adjustment (you can leave this part out, but the volume will always be at the maximum) and then through a 10 µF capacitor that provides *AC coupling*, so that the audio waveform is centered at 0V rather than 1.65V (one-half the Trinket's operating voltage, V_{CC}). The output is split to both the right and left channels of a 1/8" phono jack, to which you can connect headphones or an amplified speaker.

It is very important that you use a 3.3V Trinket (Trinket 3V) for this project, which is shown in Figure 6-55 and Figure 6-56. The Trinket 3V's digital pin voltage is directly compatible with the flash memory chip. Using a 5-volt microcontroller would require level-shifting circuitry, adding to the cost and complexity of the project.

The power source can be anything the Trinket can handle: a small 3.7V LiPo cell, three or four AA or AAA alkaline cells, etc. You can also plug it into USB, but there is a 10-second timeout while the bootloader runs its course before the playback sketch runs.

You *must* disconnect Trinket pin #4 (audio output) before you upload code to the board! If you solder the circuit permanently in a proto-board, it is strongly recommended that the Trinket be socketed with two five-pin pieces of female header. You could add a jumper between pin #4 and the RC filter instead so you can disconnect it when you need to upload new code to the chip, but I recommend socketing the Trinket with a female header.

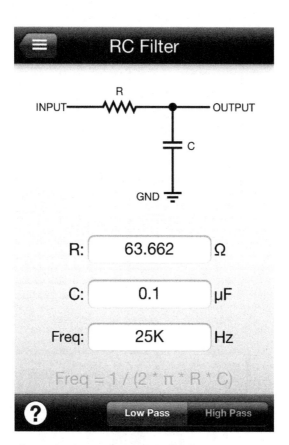

Figure 6-54. *Calculating the values for the RC filter*

 The audio connection interferes with USB. Disconnect pin #4 or unplug the Trinket from the breadboard before uploading code, then reconnect it afterward.

Load the sound playback sketch in the Arduino IDE: File→Sketch-book→Libraries→Adafruit_TinyFlash→TrinketPlayer.

Select Adafruit Trinket 8 MHz from the Tools→Board menu. Disconnect Trinket pin #4 (or remove the Trinket from the circuit), press the reset button, then click the upload button in the IDE. After uploading, assuming all else is wired properly, your audio should start playing immediately.

This code works only on the Trinket. It uses special registers and will not compile on the Uno or other Arduino boards.

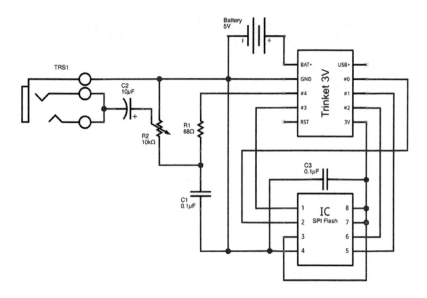

Figure 6-55. *Schematic for the audio-playing circuit*

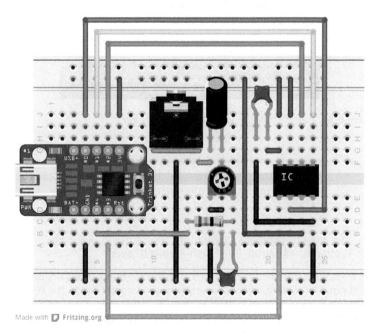

Figure 6-56. *The breadboard layout for the audio player*

Use

As the code is currently written, the sound will loop forever. You could change this to stop after the music plays, then use the reset button to restart.

This project begs for a nice little package like a mint tin. You may select any appropriate packaging, as the circuit is very compact.

Now we can ask the question: why couldn't the Trinket's EEPROM be used? Compared to the Winbond chip, at 1 megabyte, the Trinket's EEPROM is only 512 bytes—too small to hold even a tiny sound snippet.

EEPROM Memory

EEPROM is a handy, nonvolatile storage space that works well for storing data such as values that are not practical to hard code into program memory (such as any data you want to read after powering off a project).

It is not practical to use EEPROM to offload RAM data, but it is mentioned here for completeness. Using EEPROM requires that you include the EEPROM library. The EEPROM library gives us two functions, `EEPROM.read` and `EEPROM.write`:

```
#include <EEPROM.h>

int EEPROM-Address;   // An unsigned value from 0 to 511
uint8_t value;        // Byte value to read or write

value = EEPROM.read(EEPROM-Address); // Read a byte from the specified
                                     // EEPROM address

EEPROM.write(EEPROM-Address, value); // Write a byte to the specified
                                     // EEPROM address
```

Although reads are unlimited, there are a finite number of write cycles—typically about 100,000—before a specific location may wear out.

An excellent example of using this library is in the Trinket Secret Knock tutorial at *http://bit.ly/knock-activated_drawer*. The Arduino reference page is at *http://arduino.cc/en/Reference/EEPROM*.

Conclusion

The projects presented in this book cover the majority of functions available on the Trinket and the ATtiny85 processor. There is a world of possible uses for such a flexible part, though. I hope the input and output methods presented in the text and these projects have helped demonstrate how to extend the possible uses for the Trinket. Read on for some final ideas on how to use this versatile microprocessor.

7/Going Further with Trinket

This chapter rounds out your knowledge of the Trinket and its ATtiny85 microcontroller.

Microcontrollers: Smaller Versus Larger

This book demonstrates the use of the Trinket in over a dozen projects, and creative people find hundreds more uses for it every week. So should the Trinket be recommend the Trinket for all projects? Absolutely not. The Trinket is perfect where it will work effectively: in small-size, low-power projects. For projects requiring additional speed, memory, or digital pins, there are other microcontrollers to consider. Changing platforms often comes at a cost, though: the Trinket is inexpensive compared to other controllers with better hardware. For a hobbyist or manufacturer, cost is often an issue.

So which other processors might you consider? The number of products is so vast and ever-changing that there is no easy answer. To give you an idea of some of the alternatives, Table 7-1 compares the Trinket to three other small microcontroller boards.

This is quite a range of products. The Trinket is the smallest board, although the Teensy is not a whole lot larger (the Uno is huge and expensive in comparison). One issue with the Teensy is it does not use an Atmel processor, although many of the functions and some libraries are compatible or have been rewritten to function on the Teensy.

Table 7-1. *Trinket versus other microcontrollers*

	Adafruit Trinket	Arduino Uno	Teensy 3.1	Arduino Micro
Pins (digital/analog)	5/3 shared	13/6	34/21 shared	20/12
PWM pins	3	5	12	7
Voltage	3.3 or 5V	5V	3.3V (5V-tolerant)	5V

Memory (Flash/RAM/ EEPROM)	8 KB/512 bytes/512 bytes	32 KB/2 KB/1 KB	256 KB/64 KB/2 KB	32 KB/2.5 KB/1 KB
Size (mm)	31 × 15.5 × 5	75.14 × 53.51 × 15.08	43.18 × 17.78	48 × 18
Approximate cost	$6.95	$29.95	$19.95	$22.95

No one processor provides the functionality needed for all projects. Rather than use a Trinket for everything, it is best if you weigh the project requirements against the capabilities of the parts available (and the budget) to make an informed decision.

The Trinket Bootloader

Let's revisit the memory map discussed in Chapter 1. See Figure 7-1.

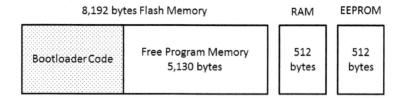

Figure 7-1. *The Trinket memory map revisited*

The ATtiny85 has 8,192 bytes of flash memory for programs, but not all of this is available for user programs. This is due to special code written both to communicate with a host computer via USB and to load programs from the host into flash. As you might recall, this special code is the bootloader. The bootloader is rather handy: you won't need a dedicated microcontroller programmer, such as those used by hardware engineers, to load programs.

However, the space used by the bootloader cannot be used by your programs. The bootloader might be needed to reload the user program due to changes. There is no way to tell the Trinket, "I am positive my program is debugged, let me use all the available space." This is mainly because the USB communication code takes up most of the reserved area—the bootloader is required to load the user's program.

It is possible to program your own (bare) ATtiny85, or bump Adafruit's bootloader out using an external programmer. First, a bit more on the Adafruit bootloader's design.

The Bootloader Design

One of the challenges Adafruit had with the Trinket design is that it wanted to have a built-in bootloader that communicated via the USB serial bus. The ATtiny85 does not have built-in USB hardware to do this like the FTDI Friend or Arduino Uno. There are existing USB bootloaders that work on the ATtiny85, but they use other companies' USB vendor identification and product identification numbers (VID and PID). Because it is not permitted by the USB Implementers Forum (*http://www.usb.org/about?*) to use others' VID/PIDs, Adafruit adapted one of the existing bootloaders, V-USB, in order to use their USB identification numbers.

V-USB is a bootloader for Atmel AVR processors produced by Objective Development Software GmbH. It is a software-only implementation of a low-speed USB device, making it possible to build USB hardware for AVR microcontrollers without requiring additional chips. V-USB may be licensed freely under the GNU General Public License, or alternatively under a non-free license.

Adafruit did not wish to distribute custom versions of avrdude or the Arduino IDE, changed to work with a new USB device (no change comes fast to the Arduino IDE). Instead, Adafruit created a USB bootloader that combines the elegance of V-USB with the well-supported and tested nature of USBtinyISP. USBtinyISP is an AVR chip programmer designed by Adafruit several years ago. Adafruit purchased an official USB Forum VID/PID before introducing the product (the identification codes are rather expensive). As the USBtinyISP is an established product (several years old), support for it is already built into the Arduino IDE.

The Trinket bootloader looks just like a USBtinyISP to a host computer. The Trinket uses the unique Adafruit VID/PID that was added to avrdude long ago; it works with only minimal configuration changes.

This answers the question posed back in Chapter 2 about why the Trinket uses a Windows driver for a device called a USBtinyISP. Adafruit's clever programming saved them from the fate of other developers who must provide extensive customizations to Arduino programming code to support their products. There are still changes required to add ATtiny/Trinket configuration support to the Arduino IDE, but they are less extensive than they might have been otherwise.

Bootloader Code

Other AVR processors (such as the one used by the Arduino Uno) have extra hardware to protect the bootloader area in flash. However, the ATtiny85 does not have a protected bootloader section in flash memory. This means it is possible to accidentally overwrite or corrupt the Trinket bootloader, preventing the loading of code. This may happen:

- If you unplug the Trinket while uploading
- If you apply voltages to the pins beyond what they are allowed to be connected to
- Random acts of magic (seriously, although I have never corrupted a Trinket once throughout many projects!)

You can use an Arduino Uno to reprogram the bootloader on your Trinket (or Gemma) if necessary.

Repairing the Trinket Bootloader

Reprogramming the ATtiny bootloader on a Trinket is possible. This may be necessary if the bootloader is corrupted, or you may want to add custom code.

 If you modify the Adafruit bootloader, the Trinket may no longer be eligible for customer support.

You can use an Arduino Uno to restore the bootloader binary code on a Trinket (or a Gemma). This loader method has not been certified by Adafruit to work with types of Arduino other than the Uno.

Make the following connections between a Trinket and the Uno:

- Trinket VBAT pin to Arduino 5V (or just power it via a battery or USB cable)
- Trinket GND pin to Arduino GND
- Trinket RST pin to Arduino #10
- Trinket pin #0 to Arduino #11
- Trinket pin #1 to Arduino #12
- Trinket pin #2 to Arduino #13

On a Gemma, alligator clips clipped carefully on the pin pads work well. The RST pin is underneath the Mini-USB jack. You may have to solder a wire there temporarily. Alternatively, sometimes you can just hold the reset button down while running the sketch (type G to start), and it might work. Soldering a wire works best, though.

Next, download the bootloader repair software from *http://bit.ly/repairing_bootloader*. Uncompress the software and run the Trinketloader sketch in the Arduino IDE.

Open up the Arduino IDE serial console (Tools→Serial Monitor) with a communication setting of 9,600 baud. When the program tells you do so, press the miniature reset button on the Trinket (or Gemma), or type the letter G into the serial console and click the send button in the upper right corner of the dialog window. You should see something similar to the screen in Figure 7-2, listing the fuses, firmware burn, and verification status. It takes about two seconds.

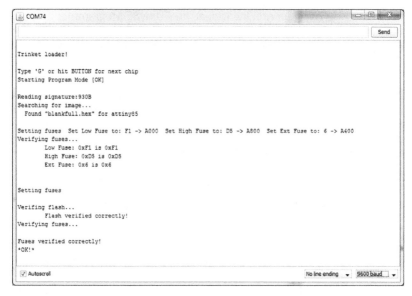

Figure 7-2. *Reprogramming the Trinket bootloader*

You can now test the Trinket with the simple Blink sketch from Chapter 3 to ensure it is working.

Programming Bare ATtiny85 Chips

You can use several types of devices to program the ATtiny85 or other AVR chips:

- An Arduino Uno
- The USBtinyISP AVR programmer, Adafruit #46
- An Arduino shield that is wired to program AVR chips, such as Adafruit #462 (this one has a zero insertion force socket to make it painless to insert and remove the programmed processor)

To program ATtiny85 chips, you need to ensure the connections are for the ATtiny85 and not the more common ATmega328P used on the Uno. See the ATtiny85 pin diagram in Chapter 1 for the right connections).

The method of using an Arduino Uno as an ATtiny85 programmer is documented in an excellent tutorial by High-Low Tech (*http://bit.ly/ATtiny85_programmer*). An Instructable tutorial that uses the High-Low technique is at *http://bit.ly/ATtiny_with_Arduino* with pictures. The Arduino IDE example program ArduinoISP is used to perform the programming.

An oft-asked question is whether you can use the source code for the Trinket bootloader for your own ATtiny85 project. It is open source: you can make a clone but cannot distribute it due to Adafruit's ownership of the device IDs (which Adafruit can't let you use due to USB Implementers Forum licensing restrictions).

The code is posted on GitHub (*https://github.com/adafruit/Adafruit-Trinket-Gemma-Bootloader*), including a precompiled *hex file* that is the binary image for the bootloader. If you corrupt your Trinket or Gemma, Adafruit states you are free to use the code to restore your product.

GemmaBoot, Adafruit's bootloader software, is free software: you can redistribute it, modify it, or both under the terms of the GNU Lesser General Public License as published by the Free Software Foundation (version 3 of the License or later).

However, Adafruit is adamant that you cannot use its VID/PID USB identification numbers for your own project. You must purchase your own identification numbers at *http://www.usb.org/developers/vendor/* if you use their code.

And there is a dilemma—once you have your own VID/PID codes (significantly denting your wallet), computers will not know that the device is the same as a USBtinyISP. To be able to program their new devices, users have recompiled avrdude, the low-level command-line program the Arduino IDE uses, to recognize other VID/PID values. But now you're back to Adafruit's original conundrum of having to modify Arduino software to work with your new chips.

If you are an advanced programmer, there might be something useful here. If you are looking to start a Kickstarter campaign for a cheap Trinket clone, perhaps another project might be better.

Other AVR Programming Methods

If you do not have an Arduino, you may have a Raspberry Pi. There is no project to date that demonstrates how a Pi may program an ATtiny85. However, a very similar project has been published on Instructables (*http://www.instructables.com/id/Programming-the-ATtiny85-from-Raspberry-Pi/*) if you wish to use it as a basis for an AVR programmer.

Also, one Adafruit user has used a Bus Pirate to reprogram a Trinket (*https://forums.adafruit.com/viewtopic.php?f=52&t=53031*).

Atmel provides free software, Atmel Studio, to program its microcontrollers. Studio or other software, such as the Eclipse environment, is often used to program ATtiny85 chips. There is additional Atmel documentation at *http://www.atmel.com/devices/attiny85.aspx*.

Code written for Atmel Studio looks more like a standard C program than Arduino code. There will be a `main` function that all C programmers are familiar with, which is hidden by the Arduino IDE. Libraries for various functions are different, and libraries written for the Arduino IDE do not necessarily work for Atmel Studio. With some skill, code may be rewritten between the two. This erects a skill barrier that was eliminated when the Arduino environment came about: the code complexity was reduced to allow more freedom to focus on the project, and not worry about perfect coding.

If you believe you would like to jump to this level, see Appendix C for more information.

The Digistump Digispark

In researching ATtiny85 development boards, you may come across a slightly older cousin to the Trinket in the form of the Digistump Digispark. The Digispark was developed as a successful Kickstarter campaign. Both have the same processor, have voltage regulation, and are programmed via USB. Table 7-2 compares board features.

Table 7-2. *Comparison of the Adafruit Trinket and Digistump Digispark*

	Adafruit Trinket	Digistump Digispark
Pins (digital/analog)	5/3 shared	6/4 shared
PWM pins	3	3
I/O voltage	3.3 or 5V	5V
Memory (flash/RAM/ EEPROM)	8 KB/512 bytes/512 bytes	8 KB/512 bytes/512 bytes
Program flash available	5,130 bytes	About 6,000 bytes

Regulated voltage current	150 milliamp	500 milliamp
Size (mm)	31 × 15.5 × 5	26.82 × 19.02 × 4.58 (11.71 with F header)
Add-on boards	Via wires or headers	Via wires, headers, or custom "shields"
USB connection	Female Mini-B	Onboard male A
Approximate cost	$6.95	$8.95

As you can see, they are very similar. One notable difference is that the Digispark uses the reset as an input/output (I/O) pin, while the Trinket exposes it as the bootloader reset. Another is that the Digispark board must be placed in a USB port or have a male A to female A extension cable, while the Trinket uses the common male A to Mini-B cable. Also, the Trinket comes in a 3.3-volt version as well as a 5-volt version.

The Trinket has its roots in the wearables market, whereas the Digispark comes from the hobbyist effort to shrink the Arduino Uno. This has defined the designs chosen by each manufacturer.

The Digispark has a number of add-on boards that Digistump calls shields, although the design is very different from that of an Arduino shield. Digispark shields may stack to obtain different functions, although this may make it awkward to program the Digispark via USB. Digistump and its user community have developed a number of libraries to provide functionality similar to that in the Arduino IDE (and similar to what we saw for the Trinket in Chapter 4). Some Digispark libraries may be compatible with Trinket, or work with minor modifications.

Advanced users have placed the Digispark bootloader onto a Trinket. The Trinket forum has this information in various discussion threads at *http://bit.ly/Trinket_forum*. The Digispark bootloader may provide a Trinket with features not found in the Adafruit bootloader, but changing the bootloader will void the Adafruit warranty.

For more information on the Digispark, see the Digistump wiki (*http://digistump.com/wiki/digispark*).

Community Resources

You cannot overestimate the value of having the Internet to provide resources to help you with your projects in general, and working with the ATtiny85 and Trinket specifically. In "the old days," all you had was paper documentation from the company, specification (spec) sheets for chips, and maybe a *Popular Electronics* article to help you. Today, companies are building successful business models on an ecosphere of customer support and tutorials using their products. Open source and user sharing provide opportunities orders of magnitude beyond those available a few years ago.

Although I recommend that you keep a copy of this book on your bench and another under your pillow, you will also be able to obtain a wealth of information on the Internet, both technical and social. For a list of other print books on Maker topics, see Appendix C.

Learning Arduino

This book assumes you have some familiarity with Arduino products, C/C++ programming, and other skills. Most documentation on learning Arduino assumes you're using an Arduino Uno or compatible board. Some users might consider learning with an Uno before tackling the Trinket, but this is not required. Here are some resources that may help with introductory topics:

Make: Getting Started with Arduino, *Second Edition (http://www.make rshed.com/Getting_Started_with_Arduino_2nd_Edition_p/ 9781449309879-p.htm), by Massimo Banzi*
An introduction to the Arduino platform, by its cocreator.

Learn Arduino *(http://learn.adafruit.com/category/learn-arduino), by Simon Monk and Limor "Ladyada" Fried*
Adafruit's Arduino guide. Start with Lesson 0.

Getting Started with Arduino *(http://arduino.cc/en/Guide/HomePage), by the Arduino Team*
An introduction to Arduino by its creators. See also the Arduino Playground *(http://playground.arduino.cc/)*.

Additional books and resources are listed in Appendix C.

Commercial Resources

Adafruit Industries *(http://www.adafruit.com)*, the maker of the Trinket, provides specifications, code libraries, and a growing number of tutorials on creative ways to use a Trinket. Here are some of the resources it provides to assist its customers:

The Adafruit main site
Many of the components used in the projects in this book are available direct from Adafruit, including the Trinket itself (http://bit.ly/ Adafruit_Trinkets).

Tutorials
The main tutorial and technical reference for the Trinket is located at *http://bit.ly/Trinket_intro*. Also see both *http://learn.adafruit.com* and *http://learn.adafruit.com/category/trinket* for Trinket-related tutorials.

Adafruit's blog (http://www.adafruit.com/blog/)
All posts tagged with "Trinket" are at *http://www.adafruit.com/blog/category/trinket/*.

Adafruit's Trinket forum (http://bit.ly/Trinket_forum)
The forum contains a listing of all questions and answers related to the Trinket to date, including any beta drivers. You can also mine the Adafruit Wearables forum, as the Gemma shares the Trinket's code base.

Digistump (*http://digistump.com/*) is the maker of the Digispark board. It provides the following resources:

Digispark libraries (https://github.com/digistump/DigisparkArduinoIntegration/tree/master/libraries)
These may not work out of the box on a Trinket, although they are more likely to take into account the ATtiny85 architecture, so they may provide a starting point if you need one.

Technical Resources

Adafruit's product pages and Trinket forum, listed in the previous section, contain information developed after this book's publication. Other resources that are available include the following:

Atmel Corporation website (http://www.atmel.com/)
Atmel makes the microcontroller in the Trinket (ATtiny85) and many, many other AVR microcontrollers. A number of resources specific to the ATtiny85 and other information on general AVR microprocessor topics are available on the company's website.

ATtiny85 data sheet (http://www.atmel.com/Images/Atmel-2586-AVR-8-bit-Microcontroller-ATtiny25-ATtiny45-ATtiny85_Datasheet.pdf)
This is the ultimate reference on how to use this chip. Warning: it's 234 pages long!

 A complete listing of ATtiny85 resources can be found at *http://www.atmel.com/devices/attiny85.aspx*.

Third-Party Sites

- Frank Zhao is the programmer of the Trinket bootloader. He has released several tutorials for Trinket on his site, *http://eleccelerator.com*

- If you need a 3D rendering of the Trinket, see the model by Gavin G. Stewart at *http://bit.ly/3D_Trinket*.

- Nick Gammon's website (*http://www.gammon.com.au/scripts/ forum.php?bbtopic_id=123*) offers a wealth of information about the ATtiny85 and other AVR microcontrollers, with well-written code examples. Nick is also the author of the SendOnlySoftwareSerial library (*http://forum.arduino.cc/index.php?topic=112013.0*).

- A good step-by-step guide on uploading code to the Trinket and Gemma is available at *http://bit.ly/Gemma_code_upload*.

Social Media Resources

Adafruit on Google+
  Adafruit posts news and information on their Google+ page (*https:// plus.google.com/+adafruit*). They have millions of followers!

Adafruit's "Show and Tell" program
  This is broadcast live Wednesdays at 7:30 p.m. US Eastern time on the Adafruit Google+ page. Many Trinket projects debut on Show and Tell.

The author's Google+ and blog pages
  For Trinket and other information, see *https://plus.google.com/+Mike-Barela*; *http://21stdigitalhome.blogspot.com* also contains Trinket-related and other information.

 You can search Google+ for the keyword #trinket to find Trinket projects and other unrelated items people tag with the term.

8/Troubleshooting

Running into problems and solving them is a defining part of the Maker experience. This chapter will help you resolve many common issues you may face when working with the Trinket.

Most issues fall into the following categories:

- Cable issues
- Connectivity issues
- Arduino IDE issues
- Common library problems
- Error messages
- Usage issues
- Manufacturer support

Your USB Cable

 Half of Trinket issues are ultimately traced to a bad USB cable or a power issue.

Very often, to get an older USB Mini-B cable, you'll scrounge in your box of old cables to find something that works. This may not get you the reliable cable you expected. Problems you may encounter include:

- Many USB Mini-B cables only have power wires and no signal wires (they were designed for charging devices only).
- The cable connections are broken or intermittent, due to flexing (often at one end).

- The wire gauge of the cable is insufficient (especially common with smaller or more inexpensive cables).
- A connector is cracked, dirty, or broken.

You may think, "This cable works for my phone, it should be good." However, the phone may not use the data wires per standard USB specifications, or it may only have power wires. That the cable works for your phone is not a sufficient indication that the cable will work for your Trinket projects.

Troubleshooting:

1. Check your connections and USB port to make sure that everything connects well.
2. If there is a problem, try swapping the cable for a thicker, more substantial one, or consider purchasing a new one.
3. As a final check, disconnect the USB cable and connect the Trinket to external power. Connect 3.7 to 9 volts, with the positive to the BAT+ pin and negative to the GND pin. If the green LED glows and you have a dim red LED for 1–2 seconds, your Trinket is working, so you are having problems with USB power. Do not expect the red LED to flash brightly for 10 seconds (bootloader mode) only on battery power: the bootloader will only act that way if the USB data lines appear like a USB port in addition to the cable supplying good power.

Buying a good, substantial cable (Adafruit #260 or similar) from a local shop or reputable online supplier will remedy many issues.

Connectivity Issues

The Trinket works best with specific hardware. Known problems with connectivity include:

1. Intermittent Trinket communications on USB 3 ports on computers (USB 3 connectors have blue plastic inside them)
2. Incompatibility issues on USB ports on some versions of the Linux operating system
3. Differences in how the Trinket communicates when compared to other Arduino-compatible devices such as the Uno
4. The lack of a serial monitor function on the Trinket

Connectivity problems are associated with specific error messages or problems. The following issues illustrate frequently encountered situations:

General communications: Is your USB cable connected to a USB 3 port?
Reconnect your Trinket to a USB 2 port. The timing in the V-USB boot-
loader may have issues with some USB 3 chipsets. If you do not have a
USB 2 port and you have issues, obtain an inexpensive USB 2
expander/hub. A powered hub is best, as shown in Figure 3-4. Plug a
USB cable from the hub into the Trinket.

I connect my Trinket via a USB cable on Windows and hear the operating
system connect, and then 10 seconds later, I hear the disconnect sound. Did
it fail?
This is the Trinket's normal operation. Unlike Arduinos, the Trinket only
makes a USB connection while it is in bootloader mode. The bootloader runs when initially plugged in and when the reset button (or pin)
is activated, and the bootloader is active for 10 seconds. The red LED
should flash during the bootloader active period, although some problems might cause the light to be dimmer than it should be (more on
that later).

I get the green power LED and a normal red flashing LED on reset, but the
Trinket appears to not communicate in any way; my program is not loaded.
Be sure you click the upload button on the Arduino IDE (the circle with
the right arrow in it) after you press the reset button on the Trinket.

I plug my Trinket into the USB cable and I have a dim red light for a short
time, but no pulsing.
This is due to poor USB cable data line connectivity. Use a known,
good cable. Also ensure there are no connections to Trinket pin #3 or
#4 during programming.

I connect the Trinket via USB and I see the green LED come on. But the red
LED will not come on at all when plugged in or when the reset button is
activated.
1. If you can, remove the Trinket from the circuit to program it. Some
   connections may interfere with programming—especially any connections on pins #3 and #4, as these are shared with the USB
   connection.

2. A remote possibility—if you have recently programmed the Trinket
   or have rewired your circuit, the bootloader may have been corrupted. See "Repairing the Trinket Bootloader" on page 202 for
   how to correct this.

3. The Trinket could be defective. Go through this entire chapter to
   review how the Trinket is used. If you still have issues after troubleshooting, see the last section on Adafruit customer support.

I get many errors when I try to upload a program in the Arduino IDE.
Make sure you've selected the correct Trinket in the Tools→Board
menu item and have selected the programmer USBtinyISP in

Tools→Programmer. Other settings do not work. If you switch back to another Arduino-compatible board, change the settings appropriately. To reprogram the Trinket, reselect the correct board and programmer. For undefined variable error messages, it may be a library problem, as discussed in Chapter 4.

I cannot find the Trinket in the list of devices in the operating system.
Depending on your operating system, you may or may not "see" it in the device listings. You will never see the Trinket as a serial device, because the USB is not actively connected the entire time. For Windows, you should see the driver in the Windows Control Panel, under Device Manager and "libusb-win32 devices" (as "Trinket" or "USBtiny"). If it is not there, ensure the Trinket is connected to the computer via a USB cable and the reset button is pressed and released. If it is there, the USBtinyISP driver has been installed correctly. It will not have a "COM" (serial) port attached (under "Properties"); this is normal. If a question mark is on the icon, you may need to reinstall the driver. See Chapter 2 for more information. For Macs, see Figure 8-1.

There are driver issues while using VMware or other virtual machine programs.
The VMware Workstation Server service is known to interfere with the USBtiny driver used by the Trinket on Windows. Stop the VMware service, if possible, to use Trinket, and then restart the service for normal VMware operation. Type "Services" in the Windows Control Panel search box to show the Services Manager and running services. This is rather complex, as stopping the wrong service can cause unpredictable results, so it's best done by the expert who installed VMware in the first place.

I cannot see the Trinket using the serial port or serial monitor in the Arduino IDE.
This is normal. Trinket on Windows communicates through the USBtinyISP driver and not as a COM or serial port. With other operating systems, the serial connection may be active only during bootloading. The serial monitor is not available, as the Trinket does not have a USB communication chip. See Chapter 3 on connectivity; this is similar to the serial monitor using an FTDI Friend and the SoftwareSerial library.

My Trinket worked when I first got it, but it is acting up now. What could be the problem?
First, check your power connections; if they are not good, correct them. Next, your circuit could be electrically problematic or miswired. If your Trinket is removable, remove it and try to load the Blink sketch for Trinket from Chapter 3 to test it out. If it works outside your project, check your project connections. If you connected to Trinket pins #3, #4, or both, these need to be disconnected when programming the Trinket, then they can be reconnected after programming.

I power my Trinket via a USB cable and it does nothing—no lights, nothing!
Does your circuit have the power connected, and to the correct pins?
Check your USB cable. Not all USB cables have all four wires (two for
power, two for data) required by the Trinket and most other USB devices. If your circuit requires more power than the power supply or USB
connection can supply, you may have severe issues. NeoPixels especially require lots of power (see Chapter 5 for calculating NeoPixel
power requirements), so do not power more than a handful from the
5V pin. You can power NeoPixels from a separate supply than the Trinket if necessary, if the ground lines are also connected.

I use Linux and have problems.
See Chapter 2 about Trinket and Linux.

*Can I charge a rechargeable battery (LiPo) connected to BAT via USB or
another mechanism?*
You can use a LiPo battery for a Trinket (a 3.7-volt LiPo is great for the
Trinket 3V), but the Trinket cannot charge the LiPo directly. You can
add a LiPo charging board such as Adafruit #259 or #280; then it will
run on the battery, and you can charge via the LiPo charging board
USB port when needed. Use a board like Adafruit #1304 for out-of-
circuit battery charging.

Arduino IDE Issues

At this point you have gone through the connectivity issues, and everything
seems to be working electrically. You appear to be having errors in the
Arduino IDE, either during the compile/verify stage or during upload.

Mac

*If I try to open the Mac version of the Arduino IDE from Adafruit.com, the
operating system says the file is "damaged, corrupt, and needs to be
trashed."*
The preconfigured download on the Adafruit site was not done by a
"signed developer" and so is trapped by OS X security. If you are using
OS X Mavericks or later, you need to update the security setting to
permit running the Arduino IDE. Go to Security & Privacy in System
Preferences, click the lock icon, and log in. Change "Allow apps downloaded from" to "Anywhere." Set it back once the preconfigured IDE
launches.

I cannot see the Trinket in the Mac USB Device Tree (or as a device in /dev).
The Trinket does not emulate a serial port to communicate like some
Arduino compatibles. You should be able to see it in a System Report,
as shown in Figure 8-1.

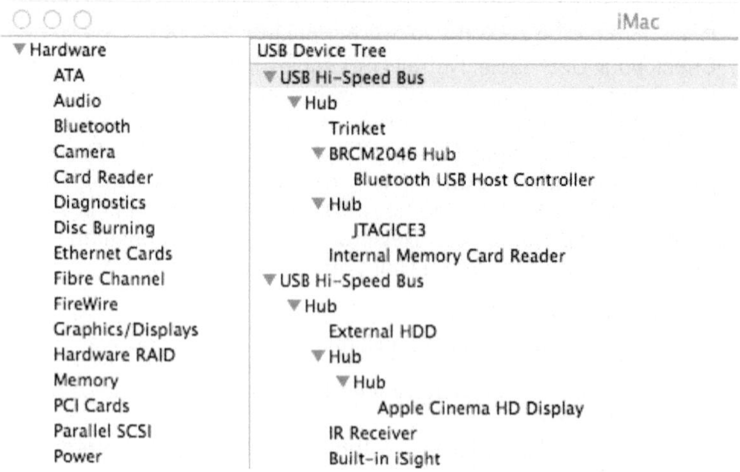

```
○ ○ ○                                              iMac
▼ Hardware               │ USB Device Tree
    ATA                  │  ▼ USB Hi-Speed Bus
    Audio                │     ▼ Hub
    Bluetooth            │          Trinket
    Camera               │       ▼ BRCM2046 Hub
    Card Reader          │           Bluetooth USB Host Controller
    Diagnostics          │       ▼ Hub
    Disc Burning         │          JTAGICE3
    Ethernet Cards       │          Internal Memory Card Reader
    Fibre Channel        │  ▼ USB Hi-Speed Bus
    FireWire             │     ▼ Hub
    Graphics/Displays    │          External HDD
    Hardware RAID        │       ▼ Hub
    Memory               │        ▼ Hub
    PCI Cards            │             Apple Cinema HD Display
    Parallel SCSI        │       IR Receiver
    Power                │       Built-in iSight
```

Figure 8-1. *A Mac OSX System Report screen showing a connected Trinket*

I get the following error message when trying to upload from my Mac: "avr-dude:error:usbtiny_transmit: usb_control_msg(DeviceRequestTO): unknown error avrdude: initialization failed, rc=-1 Double check connections and try again, or use -F to override this check."

This is seen with a Mac using a wireless mouse with a USB transceiver. Unplug the mouse transceiver USB connection and use another mouse (a trackpad or wired mouse) to program the Trinket. This error may also indicate a conflict between the Arduino IDE you are using and another version of the IDE on the same machine. Ensure all versions of avrdude are updated to the latest version on your machine.

I have the error "no connection to an IOService (expected 4, got -6)..."

See "Upload Errors" on page 221 to change the chip erase delay. This happens more often on older computers. Also consider changing the USB port, again with USB 2 preferable over USB 3.

I have VMware on my Mac and have USB issues.

You should consider stopping the VMware services. See the Adafruit Trinket forums for other solutions if you have to use VMware and have problems.

I get a "parallel port not available" message on my Mac.

If an original *avrdude.conf* file is found without Trinket changes, it often has references to parallel port programmers for avrdude that OS X does not like. Remove all references to parallel port programmers, or download the confirmed Mac *avrdude.conf* from Step 2 of Adafruit's instructions on setting up the Arduino IDE (*http://bit.ly/Arduino_IDE*).

Common Library Problems

There are many problems you can have with libraries; Figure 8-2 shows one example.

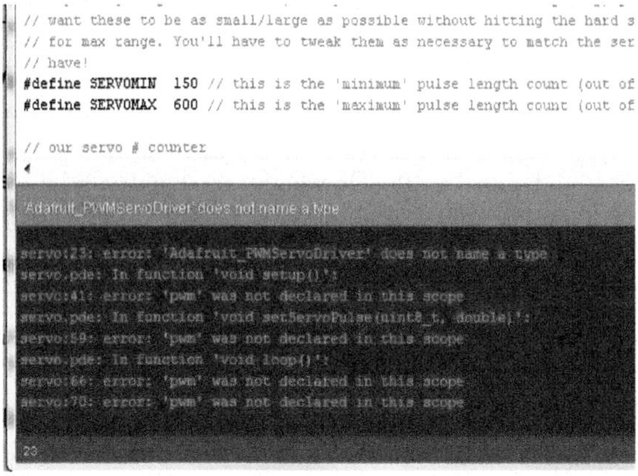

Figure 8-2. *A sample of a library-related error message*

The most common library-related error messages take the form "XXXX does not name a type" or "YYYY not declared in this scope." They mean that the compiler could not find the library. This can be due to any of the following causes:

The library is not installed
 See the steps in Chapter 4 to install a library correctly.

Wrong folder location
 The IDE will find standard libraries and libraries installed only in the sketchbook *Libraries* folder. It will not be able to find libraries installed elsewhere.

 The specific library folder must be at the top level of the *Libraries* folder. If you put it in a subfolder, the IDE will not find it.

You do not have a "Sketchbook" folder
 It is there, but on a Windows or Mac OS X machine it is named *Arduino* (on Linux it is named *Sketchbook*). See "Where to Install Libraries" on page 42 for further details.

Incomplete library
 You must download and install the entire library. Do not omit or alter the names of any files inside the library folder.

Wrong folder name
 The IDE will not load files with certain characters in the name. Unfortunately, it does not like the dashes in the ZIP file names generated by

GitHub. When you unzip the file, rename the folder so that it does not contain any illegal characters. Simply replacing each dash (-) with an underscore (_) usually works. If the folder has the word "master" on the end (usually preceded by a dash), remove that also. The best method to see what the library name should be is to look at the sample code to see what the sample expects the library name to look like.

Wrong library name
The name specified in the `#include` line of your sketch must match exactly (including capitalization!) the class name in the library. If it does not match exactly, the IDE will not be able to find it. The example sketches included with the library will have the correct spelling. Just cut and paste from there to avoid typographical errors.

Multiple versions
If you have multiple versions of a library, the IDE will try to load all of them. This will result in compiler errors. It is not enough to simply rename the library folder; it must be moved outside of the sketchbook *Libraries* folder so the IDE will not try to load it.

Library dependencies
Some libraries are dependent on other libraries. For example, most of the Adafruit graphic display libraries are dependent on the Adafruit-GFX library. You must have the GFX library installed to use the dependent libraries. This is also true for libraries that use I^2C that also expect the Wire library, which for the Trinket is TinyWireM.

"Core" libraries
Some libraries cannot be used directly. The Adafruit-GFX library is a good example of this. It provides core graphics functionality for many Adafruit displays, but cannot be used without a specific driver library for the display you are using.

Forgetting to restart the IDE
The IDE searches for libraries at startup. You must shut down *all* running copies of the IDE and restart before it will recognize a newly installed library.

Here are some other library questions you might have:

I found a wonderful Arduino library that does what I need, but when I try to use it on my Trinket, I get errors. What can I do?
First, check Chapter 4 to see if the library or functionality you want to use is already listed. If not, your library find was probably coded for other microcontrollers. Those libraries might use large memory spaces or additional hardware in other microcontrollers, which may not work on the Trinket. If you understand how the library code works, you may be able to fix some errors yourself. Performing a Google

search for the library name and "ATtiny85" may produce pages where others found the same circumstance and recoded the library.

Does a library I found on the Internet work with Trinket?
As there are hundreds of libraries out there written by all sorts of people, and the Arduino Uno is the common platform, there's a good likelihood it won't work. But it doesn't hurt to try—see the previous question to proceed.

I have problems using the Servo library with the infrared sensor code. Help!
The IR code uses a loop to check for infrared sensor pulses to decode them. The Servo library refreshes the servo position via interrupt code every few milliseconds, so yes, the servo refresh may interfere with the decoding. Temporarily disabling interrupts while counting pulses might work, with careful coding to not make the servos mad. Larger Arduinos typically juggle this with their beefier timers and larger IR libraries.

Error Messages

Error messages may fall into the general categories listed next.

Compilation Issues

I get a message to the effect, "A device attached to the system is not functioning. (expected 4, got -5)."
If you get only this message, your program may have loaded successfully. Adafruit states it is only a warning from the Trinket when asking for data. This may happen depending on the reset button timing or when loading a large program. See if your program is running, and if not, try to load it again. If you get this message with others, focus on the others. Also be sure you have updated your Arduino IDE to include the new *ld.exe*, `chip_erase_delay`, and latest avrdude version (these should all be included in the latest Adafruit preconfigured IDE; see Chapter 2). Finally, you might have the Trinket on a USB 3 port—try to use a USB 2 port or powered USB 2 hub.

My program will only compile a sketch up to about 4,400 bytes, not the 5,310 or so discussed in the book.
This indicates that the *ld.exe* program from the standard Arduino IDE is still being used. See Chapter 2 for details on changing the linker to a new version that will support the full amount of program memory available on the Trinket.

I get a message like "avrdude: stk500_getsync(): not in sync: resp=0x00."
There are two possibilities here. First, ensure the Arduino IDE is set to Adafruit Trinket 8 MHz, Adafruit Trinket 16 MHz, or Gemma 8 MHz, as appropriate, and not Arduino Uno or something else. Second, be sure

you press the Trinket reset button and the red LED is flashing when you upload your code (you have a short 10-second window to do so). This error generally indicates a serial read error. Unplug and replug the Trinket as a last resort.

I get an error like "'+TCCR2A+' was not declared in this scope." (or TCCR3A, TIMSK3, OCR3A, OCR2A, TCCR2B, etc.).
This is a sign that you are trying to use code that uses Timer 2 or Timer 3, available on large AVR microcontrollers but not the Trinket. Trinket has Timers 0 and 1 available, and their use is different than on the Uno. Look at the code or library to see if the timer code can be changed to Timer 0 or Timer 1. Any code requiring 16-bit timers only will not be compatible, as there are only 8-bit timers on Trinket. You can use the ATtiny85 datasheet or online resources to see about changing the timer code, but you might want to look for a different library designed to work with the ATtiny85.

I get the message "In function 'void loop()': sketch_may10a:19: error: 'A1' was not declared in this scope."
The analog pins A1, A2, and A3 should be referred to using the numbers 1, 2, or 3, respectively. The difference in the pin designating change is due to a bug in the Arduino IDE for Trinket and Gemma. An example: `analogRead(1);` reads pin A1, which is the same physical pin as digital 2. There is no confusion with the digital pins, but remember the analog and digital pin numbering are different, as shown in Figure 1-4. Also, calls to `analogWrite` use digital pin numbers, as `analogWrite` actually outputs a Pulse Width Modulation signal on certain digital pins (0, 1, and 4 with code).

I get an error like "/Applications/Adafruit Arduino.app/Contents/Resources/Java/hardware/tools/avr/bin/../lib/gcc/avr/4.3.2/../../../../avr/lib/avr25/crttn85.o:(.init9+0x2): relocation truncated to fit: R_AVR_13_PCREL against symbol `exit' defined in .fini9 section in /Applications/Adafruit Arduino.app/Contents/Resources/Java/hardware/tools/avr/bin/../lib/gcc/avr/4.3.2/avr25/libgcc.a(_exit.o)." The Internet gives instructions about linkers, etc., but it is not helpful.
If you get any error messages about PCREL or symbols, the linker has detected a program that compiles to more than 5,310 bytes. Your code is too big for Trinket. See Chapter 4 for some tips on memory optimization.

Can I write code that will compile one way for Uno, and another for Trinket?
Maybe, if the Arduino IDE internal preprocessor allows it. The defined symbols for the processor are (note each __ is two underscores):

- __AVR_ATtiny85__: ATtiny85 processor (Trinket, Gemma, Digispark)
- __AVR_ATmega328P__: Uno R3 and compatibles

- __AVR_ATmega32U4__: Leonardo, Flora, Micro, Esplora

- __AVR_ATmega2560__: Mega 2560

  Note that __AVR_ATtinyX5__ does not work; you need to replace the X with an 8.

You might think the following code could work:

```
#ifdef __AVR_ATtiny85__
#include <TinyWireM.h>   // For Trinket
#else
#include <Wire.h>        // For larger microprocessors
#endif
```

But most often the Arduino IDE will try to include *both* libraries, in an effort to be helpful—not the desired thing at all! This is why Adafruit often makes separate libraries for the Trinket/Gemma code. `#ifdef` usually works the same as `#if defined()`, but not always; this is well known and another IDE idiosyncrasy that is still not fixed.

If you wish to write code with complex preprocessor directives, you should check the Arduino forums (*http://forum.arduino.cc/*) for the latest information, as this is an evolving issue in updated Arduino IDE versions.

Upload Errors

Some users have reported intermittent IDE errors during upload even with the correct software. For some computers—perhaps slower machines—one change will make a difference.

Find the directory where the Arduino IDE you installed is located via Windows Explorer or Finder in Mac, and go to that directory. You will see a number of subdirectories that make up the IDE software. Go to the *hardware/tools/avr/etc* directory. You will need to open the file *avrdude.conf* with a text editor (Notepad or WordPad for Windows; TextEdit or your favorite editor for Mac). Scroll down a long way to find the line under the **ATtiny85** heading (do not make changes under any other headings). Make changes to the one line noted here:

```
chip_erase_delay = 900000;
```

Change this line to:

```
chip_erase_delay = 400000;
```

Save the file and get back to your desktop. If you have the Arduino IDE open, close and restart it to make sure the change takes effect. If you configured your own IDE download, also ensure you have the latest avrdude files. If you have multiple versions of the Arduino IDE, avrdude, or both on your computer, it is easy to edit the wrong configuration file, so be sure you check if you are still getting errors and you believe you edited the right file.

An additional message you may encounter when compiling a program for a Trinket:

I uploaded the Blink sketch to my new Trinket after pressing the reset button. I got an "expected 4, got -5" error, and now the bootloader LED continues to blink. Is it stuck or broken?

Your Trinket is working fine and running your sketch. As the Blink sketch blinks the same pin #1 red LED that the bootloader flashes for 10 seconds, you can easily be tricked into thinking the bootloader is still running when your Blink sketch is running. This is why in Chapter 3, it is recommend that you change the `delay` function times so the LED will blink faster during your sketch compared to the bootloader. You can ignore "4, -5" errors if it appears your code is running fine. If the code appears not to run, look for other problems.

The Serial Monitor

If you use code for other Arduino processors, it may include code that assumes a serial monitor is built into the board (such as that built into the Arduino Uno and other boards). The Trinket does not have a dedicated USB chip, so it cannot perform the same way. See Chapter 5 for a discussion of serial communication. There is a method of using SoftwareSerial to act as a serial monitor but you will need an external FTDI Friend or similar board to provide a serial interface.

Usage Issues

The following issues may be encountered while using the Trinket. Often they are not errors, but rather differences in how the Trinket behaves versus other microcontrollers:

I cannot get `analogRead` to give changing values; it's like it's broken. Or I get different values for every call—what gives?

Recall from Chapter 1, analog 1 is Trinket pin #2, analog 2 is Trinket pin #4, and analog 3 is Trinket pin #3. And unlike when programming Uno, with Trinket you cannot use A1 to refer to analog 1, etc., as those mappings were not included in the definitions for the ATtiny85. However, the numbers work fine—just use the number 1 for A1, 2 for A2, or 3 for A3. If a value is not "right," ensure you have the right analog pin number for the physical pin you are connecting to. See Chapter 1 for pin mappings.

I am using `analogRead` on Trinket pin #3 (analog 3), and it does not give the same type of readings as performing reads on Trinket Pin #2 (analog 1).

If you look at the Trinket pin connection diagram in Chapter 1, you'll see that there are components on pins #3 and #4 to support USB communication. Pin #2 has no extra components. You may add additional components on pins #3 and #4 to compensate for this, or factor the effects of the values returned from `analogRead` into your code as was done in the Trinket Alarm System program in Chapter 6.

When I apply power to my project, the Trinket executes the bootloader for the first 10 seconds before executing my code. Can I skip this 10-second pause?

In general, no; this is not easily changed due to how the Trinket was designed. An advanced user could use an AVR programmer to either upload a program without the bootloader or change the code reset vector.

I would like to use a Trinket 5V. What type of power options should I consider?

A Trinket 5V generally likes 5.2 to 16 volts on the BAT terminal, but it still works down to about 3.3 volts as long as you do not expect voltage from the 5V regulated voltage pin. Please do not put two LiPo batteries in series or parallel to get more voltage or current: it could cause a fire due to melting batteries. Remember the Trinket 5V will still work reliably at 8 MHz on 3.7 volts (and there is an optional JST connector that allows this to happen more easily). If you really need a higher-voltage LiPo, the 7.4-volt units used in model vehicles may be good, although you will need the appropriate charger also. Alkaline or rechargeable AA or AAA batteries are fine; it's best to use three of those batteries for Trinket 3V (4.5 volts) or four for Trinket 5V (6 volts).

The documentation says Pulse Width Modulation (PWM) is available on Trinket pins #0, #1, and #4, but the `analogWrite` function only works when told to use pin #0 and pin #1. How do I get pin #4 to do PWM?

True, `analogWrite` does not recognize pin #4 in some older versions of the Adafruit-supplied Arduino IDE. However, there is code to use Timer 1 to do PWM on pin #4. See the Analog Meter Clock code in Chapter 5 for the two functions you can use.

I have output signals on pins and I can see them with an oscilloscope, but LEDs on the same pins will not light.

If you do not set a pin for digital output (via the `pinMode` function), the drive current will not be enough to power external circuitry even though you can see it with a high-impedance oscilloscope.

Windows complains about loading the drivers because they are unsigned.

See "The USBtinyISP Driver for Windows" on page 11 on installing drivers for Windows.

I would like to try using Linux. What are the pitfalls I need to look out for?

Your programmer may need access to the USB port, but this is controlled by root. See Chapter 2 for details. Also, the USB code in the Linux kernel may not like the Trinket timings, but there's not much you can do about that. Go ahead and experiment.

I would like to use the Trinket for a specific function; will this work?

Possibly. You will have to review the resources required (hardware: memory, pins, etc.; software: code and libraries) to accomplish what you want and make a decision. If you need a small board that offers more resources, see Chapter 7.

Will Trinket interface to the hardware I have?

The answer is similar to that for the previous question: possibly. If it takes more than five pins, probably not. You will need to do some research on the requirements for your item, or do some experimenting. Some folks have done some remarkable projects that others might have thought a stretch for Trinket. Everyone likes a surprise.

I have connected well over a hundred NeoPixels to my Trinket, per Chapter 5. I can only light about 110 or so. Is there a bug in the NeoPixel library?

No bug. The Trinket only has 512 bytes of variable memory (static RAM), and each NeoPixel takes 3 bytes of memory. Between this and the memory required to run a program, there is no additional memory for more pixels. If your project requires more than about 110 NeoPixels, you should consider another microcontroller (see Chapter 7 for some alternatives).

Are the Trinket EagleCAD circuit board (PCB) layout files available?

Yes; see *https://github.com/adafruit/Adafruit-Trinket-PCB*.

Manufacturer Support

Adafruit Industries makes customer service and satisfaction a cornerstone of its business. If you still have problems after troubleshooting, you can go to the Adafruit forum (*http://forums.adafruit.com/viewforum.php?f=52*) to describe your situation. The helpful forum moderators will be able to assist with additional troubleshooting.

There are also many tutorials on using Trinket and more at *https://learn.adafruit.com/category/trinket*.

After posting to the Adafruit forum, if it is evident your Trinket is defective, at their discretion, Adafruit may replace it.

A/Making Electronic Sounds

Sound is a very personal part of any project. Everyone has his particular vision of how something should sound. The Trinket Animal project in Chapter 6 used customizable sounds. This type of sound generation can be added to many Trinket projects.

The Arduino **tone** function does not work with the Arduino IDE used to program Trinket. This is not a hindrance, though, as tone generation can be done simply with a tiny bit of code.

To experiment with sounds, you can use a sound creation program on Trinket or other Arduino compatibles. The program shown in this appendix comes with several preprogrammed sounds:

- A bird chirp (best)
- Cat meows (a "meeee-ow," a "me-oooow," and a "mew")
- Dog sounds ("ruff" and "arf")
- A somewhat mechanical-sounding owl

 "Ruff" and "woof" are conventional representations in the English language of the barking of a dog. *Onomatopoeia* or imitative sounds vary in other cultures: people "hear" a dog's barks differently and represent them in their own ways. Some equivalents of words in other languages are listed in Wikipedia (*http://en.wikipedia.org/wiki/Onomatopoeia*). You can hear different sounds at *http://www.bzzzpeek.com*.

Part of the creative process is making your project sound like what you believe it should sound like. This will probably involve a fair amount of experimentation. Most advanced computers use samples of real sounds taken with a microphone. This sound data would take up too much memory for Trinket's onboard memory, though, and using a ROM memory chip takes many microcontroller pins.

The method used here turns a digital pin on and off very fast to make sounds at various frequencies. A piezo speaker is used to output the sound, although a Trinket pin could just as well drive an audio amplifier and speaker (with more power draw and a larger size).

The beep function used in Example 3-2 maps specific frequencies and durations to a digital pin. Example A-1 uses a different function, playTone, which takes fewer mathematical calculations and will work faster. But the values you send it are not in Hertz. Rather, the highest value (15000) is a low frequency, and the highest shrill is the lowest value (50). Number values beyond these are not reproduced by a piezo speaker. The specified duration is in milliseconds (a millisecond is one thousandth of a second).

If you have an Arduino Uno/Leonardo/Mega/etc. handy, you can use a serial monitor to select values for different sounds using a potentiometer and see the values on the serial monitor. If you have a Trinket, you can listen to a tone and display the output frequency by connecting an FTDI Friend as a serial monitor. The circuit to generate sounds is shown in Figure A-1. The code is listed in Example A-1, which may be downloaded from the repository for this book (*http://bit.ly/GettingStartedWithTrinket*) (directory *Appendix A Code*, subdirectory *AppendixA_01Soundtest*). If you use a Trinket, note the slight code changes required.

This code will run on an Arduino Uno, Leonardo, Mega, or other AVR-compatible microcontroller. The code also works on a Trinket with SPEAKER changed to 2 and using the lines defining the SoftwareSerial library, which communicates to an external FTDI Friend serial board.

You can add value/duration values together to make increasingly complex sounds. A tone can be ramped up or down to get effects, also.

To start, map out your desired sound into component sounds. For example, for the "meow" sound for a cat:

1. Start with the "m" sound: use the varyFrequency function to find an "mmm" (maybe 5100 on the output scale). The sound is short, so I chose 50 milliseconds.

2. The "e" is "eeee", which lasts longer, so the value 394 sounded right. Also, the sound is longer, so trial and error led to a value of 180 milliseconds.

3. The "o" is a more complex sound, starting high (990) and getting a bit lower (1022). How long you run the loop and the duration specified for each sound will vary how it sounds.

4. Finally, "w" is enough like "m" that it is repeated, as no better sound could be found.

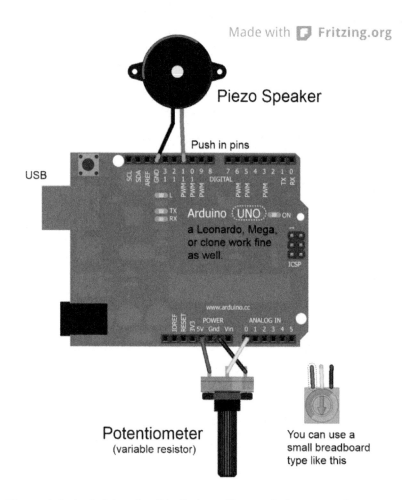

Made with ▮ Fritzing.org

Piezo Speaker

Push in pins

USB

Potentiometer
(variable resistor)

You can use a
small breadboard
type like this

Figure A-1. *An Arduino circuit to find specific sounds for your project*

Example A-1. Code for testing various sounds and sound combinations on an Uno

```
/* Trinket Sound Testing Module

   Works on Arduino Uno, Leonardo, Mega, etc.
   This can work on Trinket with pin changes and an FTDI Friend.
*/
#define SPEAKER 11 // Piezo Speaker pin (positive lead)
               //    for Trinket, change to 2
#define POT A0     // For Trinket, use 1 for pin #2 (not A1)
//#include <SoftwareSerial.h>
```

```
//SoftwareSerial Serial(1,0); // If Trinket, serial out pin #0

void setup() {
  Serial.begin(9600);
  pinMode(SPEAKER, OUTPUT); // Important to set pin as output
}

void loop() {      ❶
  chirp();  delay(2000);
  meow();   delay(2000);
  meow2();  delay(2000);
  mew();    delay(2000);
  ruff();   delay(2000);
  arf();    delay(2000);
  hoot();   delay(2000);
// varyFrequency();   ❷
// scale();           // Plays a tone scale
}

void varyFrequency() {
// Use potentiometer to produce one tone per value of pot
// Food for getting pitch value to use in making sound routines
  int reading;
  const uint8_t scale = 1; // 1 for high frequencies, scale up to 15 for
                           // lowest freqs
  reading = scale * analogRead(POT);    ❸
  playTone(reading, 1000);
  Serial.print("Freq = ");
  Serial.println(reading);
}

void chirp() {               // bird chirp
  for(uint8_t i=200; i>180; i--)
    playTone(i,9);
}

void meow() {                // cat meow (emphasis on "me")
  uint16_t i;
  playTone(5100,50);         // "m" (short)
  playTone(394,180);         // "eee" (long)
  for(i=990; i<1022; i+=2)   // vary "ooo" down
    playTone(i,8);
  playTone(5100,40);         // "w" (short)
}

void meow2() {               // cat meow (emphasis on "ow")
  uint16_t i;
  playTone(5100,55);         // "m" (short)
  playTone(394,170);         // "eee" (long)
  delay(30);                 // wait a tiny bit
  for(i=330; i<360; i+=2)    // vary "ooo" down
    playTone(i,10);
  playTone(5100,40);         // "w" (short)
}
```

```
void mew() {              // cat mew
  uint16_t i;
  playTone(5100,55);      // "m" (short)
  playTone(394,130);      // "eee" (long)
  playTone(384,35);       // "eee" (up a tiny bit on end)
  playTone(5100,40);      // "w" (short)
}

void ruff() {             // dog ruff
  uint16_t i;
  for(i=890; i<910; i+=2) // "rrr" (vary down)
    playTone(i,3);
  playTone(1664,150);     // "uuu" (hard to do)
  playTone(12200,70);     // "ff" (long, hard to do)
}

void arf() {              // dog arf
  uint16_t i;
  playTone(890,25);       // "a" (short)
  for(i=890; i<910; i+=2) // "rrr" (vary down)
    playTone(i,5);
  playTone(4545,80);      // intermediate
  playTone(12200,70);     // "ff" (shorter, hard to do)
}

void hoot() {                      // owl hoot (fairly mechanical...)
  uint16_t i;
  playTone(50,2);                  // short low to make "h" (hard to do)
  for(i=0; i<240; i=i+2) {         // "oooo" sound (vary tones slightly
    if(i%2) playTone(1832, 2);     // every 2 ms to soften)
    else playTone(1800, 2);
  }
  playTone(1790,10);               // bring up slightly near end
  playTone(14000,3);               // next 6 to simulate a "t," which is
  delay(2);                        // again a hard sound to do
  playTone(14000,3);
  delay(2);
  playTone(14000,3);
}

void playTone(uint16_t tone1, uint16_t duration) {   ❹
  if(tone1 < 50 || tone1 > 15000) return; // These do not play
  for (long i = 0; i < duration * 1000L; i += tone1 * 2) {
    digitalWrite(SPEAKER, HIGH);
    delayMicroseconds(tone1);
    digitalWrite(SPEAKER, LOW);
    delayMicroseconds(tone1);
  }
}

void scale() {                     // Play different frequencies in sequence
  for(uint16_t i=50; i<15000; i++) {
    playTone(i,20);
  }
}
```

❶ The loop function plays the different pre-made animal sounds. If you add a new sound, comment out the animal functions.

❷ Use the varyFrequency function to set a tone you might wish to use.

❸ analogRead produces values from 0 to 1023. To get tones from 0 to 2047, change SCALE to 1; to get 0 to 3071 change it to 2, etc. This provides a bit more granularity at lower frequencies.

❹ The function playTone plays a tone on a piezo speaker. Shorter values produce higher frequencies—this is the opposite of beep but avoids complex math.

That is "meeee-ow." For a cat, "me-oooow" and "mew" are variations. You may feel they do not sound cat-like enough, which is fine. You can define your purr-fect sound using these methods.

The math used in playTone is much simpler than the code used in the beep function used in the Theramin project, which is relisted here:

```
void beep (unsigned char speakerPin,
           int frequencyInHertz,
           long timeInMilliseconds) {
  int x;
  long delayAmount = (long)(1000000/frequencyInHertz);
  long loopTime = (long)((timeInMilliseconds*1000)/(delayAmount*2));
  for (x=0; x<loopTime; x++) {
    digitalWrite(speakerPin,HIGH);
    delayMicroseconds(delayAmount);
    digitalWrite(speakerPin,LOW);
    delayMicroseconds(delayAmount);
  }
}
```

The logic in the beep function uses more complex mathematical functions (two 32-bit-long divides with additional multiplications) than playTone. The 8-bit Trinket can do 32-bit integer math, at the expense of time. To keep the calculations simpler and less time-consuming, playTone sacrifices code simplicity for code speed. When playing several short sounds at different frequencies, it can be important not to spend too much time calculating.

More mechanical sounds or musical notes may also be created using these methods. If creating complex sounds (a musical phrase with different frequencies) is not needed, then the beep function, with its similarity to the Arduino built-in tone function, would be easier to use.

Sound is a powerful method of interaction with users. Trinket has the capability to generate sounds with as little as one external component. With interactive electronics projects, sound becomes a great method for providing feedback and circuit or sensor indications.

B/Parts Sourcing

The Trinket is manufactured by Adafruit Industries (*http://www.adafruit.com/*). Adafruit is typically listed as the supplier for parts in this book if they carry them, both for compatibility and ease of ordering. The liberal use of Adafruit components also stems from tutorials that originally were posted to the Adafruit Learning System (*http://learn.adafruit.com/*).

Trinkets and other Adafruit parts may be obtained from Adafruit distributors and some third-party electronics sellers. A list of official Adafruit distributors may be found at *http://www.adafruit.com/distributors/*. Not all distributors carry the full line of Adafruit parts.

One of the most frequently asked questions on forums is: "I bought this part on eBay, will it work with Trinket?" The answer is "maybe". You will probably have to try to figure things out on your own—that great price online may have come with little or no technical support. You can try Internet searches to fill the gap. The support information and forums on established vendor websites often make it worth paying a bit extra.

Trinket

The main Trinket supplier in the United States is Adafruit (*http://www.adafruit.com*), where you can buy all of the following:

- Trinket 5V (each), Adafruit #1501
- Trinket 3V (each), Adafruit #1500
- Trinket 6-pack (3 × 5V and 3 × 3V), Adafruit #1509

 Adafruit's Trinket is different from their Trinket Pro, which was released after the Trinket. The Trinket Pro is substantially different from the Trinket used in this book.

The Trinket is also available at Maker Shed (*http://www.makershed.com/search?q=trinket&type=product*) and from additional distributors (*http://www.adafruit.com/distributors/*) listed on the Adafruit website. Distributors will probably have their own part numbers. Be sure to carefully choose the type of Trinket you want to use—for some projects, the Trinket voltage is not interchangeable.

Displays

The displays used in this book rely on the Adafruit I²C backpacks.

Without a breakout board to simplify the interface, the Trinket could probably not (easily) handle the large number of pins most displays require.

Not all libraries for I²C displays have been tested with Trinket but they are probably compatible, with possible minor code changes for the ATtiny85.

One type of display you may wish to use is a serial display. These use TTL serial communication to send commands and text. SparkFun (*https://www.sparkfun.com/*) and others sell these. You can use the SoftwareSerial library to write to serial displays.

Sensors

Most sensors used in the Arduino world have libraries to help you use them. Some may be more complicated to use than others, due to command interfacing or bus communications. You can use other sensors by reading a digital or analog pin. When researching the sensor you want to use, look at the datasheet to see if the sensor can be read easily with a Trinket. If the datasheets appear rather cumbersome, some searching on the Internet may help you find a project where someone else has used that specific sensor. This is not cheating! You are just using the blood, sweat, and tears of others. This allows you to work on the project and not focus on the parts.

Resistors

You will want to source a number of resistors to work with a Trinket. They are used for many things, including LED current limiting, pull-ups, and more. Most electronics parts sellers carry resistors in single-value packs or assortments with a range of values in a larger kit. The larger kits typically provide more values per resistor and give a number of resistors you can experiment with. Adafruit does not carry through-hole resistors as of the publication of this book.

Resistors come in different wattages. Wattage is voltage multiplied by current. The maximum wattage required by a typical digital signal on an ATtiny85 pin can be calculated as follows: 5 volts × .030 amperes = .150 watts. So, for signals, quarter-watt (0.25W) resistors should be fine. If you think you will use resistors in higher-current circuits, half-watt (0.5W) should be fine (they are just a bit bigger in size).

Maker Shed (*http://www.makershed.com/*) offers assortments such as the 1/4-watt, 365-piece resistor kit (#MKEE4) and the 1/2-watt 365-piece resistor kit (#MKEE5). RadioShack (*http://www.radioshack.com/*) sells single values and assortments such as the 1/4-watt 500-piece resistor assortment (#271-312) and 1/2-watt 100-piece resistor assortment (#271-306).

Other vendors have similar assortments or single values.

Nuts and Bolts

The Trinket has mounting holes that are a good fit to M2 size screws. Metric hardware that small may not be readily available. Micro Fasteners (*http://www.microfasteners.com/*) in the US has M2 screws in various lengths. Cheese Head screws (a flat, thin head) are a good choice, but you can choose your favorite type.

Having other assorted screws on hand is always a good idea. Scavenged screws, nuts, and washers are a staple of Makers everywhere. Taking apart unneeded items can help you build up a good collection. If you need something specific that you do not have, buying some hardware is always an option.

Kits

Some suppliers have parts kits for experimenters, including some targeted to Arduino enthusiasts. The value in these kits varies, and the parts may not be the same as the ones used in this book. If you have kit parts on hand, go ahead and see about using them.

C/Publications

The following publications may assist you in working with the Trinket.

ATtiny85

ATtiny 25/45/85 Datasheet, Atmel Corporation, 2013
The official datasheet is *the* authority on how the ATtiny85 works. You might cringe at 234 pages of technical data, but there is a great deal of helpful information here: *http://bit.ly/ATtiny_Datasheet*

Books

Make: Basic Arduino Projects, *by Don Wilcher (Maker Media)*
This companion book to the Ultimate Arduino Microcontroller Pack (Maker Shed #MSUMP1) provides 30 clearly explained projects that you can build with this top-selling kit right away—including multicolored flashing lights, timers, tools for testing circuits, sound effects, motor control, and sensor devices.

Make: Electronics, *by Charles Platt (Maker Media)*
Licking a battery is your first assignment as you begin the lessons in this unique and colorful guide that uses "learning by discovery" to teach basic electronics. First, you build and experience the circuit (as in the aforementioned battery licking), then you learn the theories behind it. You will "burn things out, mess things up—that's how you learn."

Make: Learn to Solder, *by Brian Jepson, Tyler Moskowite, and Gregory Hayes (Maker Media)*
Learn the fundamentals of soldering and pick up an essential skill for building electronic gadgets. Discover how to preheat and tin your iron, make a good solder joint, desolder cleanly (when things don't go so well), and use helping hands to hold components in place.

Make: Getting Started with Arduino, *Second Edition, by Massimo Banzi, the cocreator of Arduino (Maker Media)*
This book offers a thorough introduction to the Arduino open source electronics prototyping platform that has taken the design and hobbyist world by storm. *Getting Started with Arduino* gives you lots of ideas for projects and helps you get going on them quickly.

Arduino Cookbook, *Second Edition, by Michael Margolis (Maker Media)*
This book is perfect for beginners to advanced users who want to experiment with the popular Arduino microcontroller and programming environment. There are more than 200 tips and techniques for building a variety of objects and prototypes, such as toys, detectors, and robots—as well as interactive clothing that can sense and respond to touch, sound, position, heat, and light.

Make: Getting Started with Processing, *by Casey Reas and Ben Fry (Maker Media)*
Processing was discussed briefly in Chapter 6. To learn more about this powerful programming environment that has an interface similar to the Arduino IDE, this is the book to start with.

Make: AVR Programming, *by Elliot Williams (Maker Media)*
This project-oriented book lets you start either with an AVR-powered Arduino or with a bare AVR chip and programmer.

Making Things Talk, *by Tom Igoe (Maker Media)*
This is the book that started the author's Maker adventure: a great book marrying microcontrollers and the outside world. Highly recommended.

Additional Resources

Maker's Notebook
You may use any notebook to document your making experience. Or you can buy a Maker's Notebook: *http://bit.ly/ Makers_Notebook*

Make: magazine *(http://makezine.com/)*
The frontrunner in cutting-edge projects and Maker news.

Index

Symbols

#if defined, 221
#ifdef, 221
#include, 124, 218
3D printing, 13, 60, 176

A

AC coupling, 194
Adafruit GFX library, 40, 49, 140, 218
Adafruit Industries, 13, 2, 207, 231
   Distributors, 231
Adafruit Learning System, 12, 41, 207,
   224, 231
Adafruit support forum, 224
Alarm
   Block diagram, 148
   Branches, 149
   Silent, 148
   Tripped, 148, 157
Analog Meter Clock project, 110
Analog pin
   In Arduino IDE, 25
   Location, 5
analogRead, 222
   Error, 220
analogWrite, 109, 220, 223
anim.h, 128
Annunciation, 148, 152
Arduino IDE, 9-11
   Adafruit version, 10
   Learning, 207
Arduino Micro, 199
Arduino Playground, 207
Arduino Uno, 1, 199, 203
arduino-nrf24l01 library, 41
Arduino-UsiSerial library, 41
ArduinoISP sketch, 204
Atmel Corporation, 3, 208, 235
Atmel Studio, 205
ATtiny85, 3, 37
   Pins, 5
Audacity, 188
AVR, 236

Programming, 203
avrdude, 7, 201

B

Backpack, 86
Batteries
   AA or AAA, 28, 70
   LiPo, 8, 28, 70
   Recharging, 215
Bitbang, 14, 104
Blog, Adafruit, 208
Bluefruit, 164
Bluetooth, 164-165
   4.0, 165
   Low Energy, 165
Board layout, 224
Books, 11, 235
Bootloader, 200-202, 208
   Mode, 22
   Repair, 202
   Skipping, 223
Breadboard, 17
Buffering, display, 51, 86
Burgess, Phillip, 68, 121, 185

C

Cadmium sulfide photocell, 30, 169, 171
Change interrupts, 108
chip_erase_delay, 221
Clock speed, 8
clock_prescale_set, 96
Color Organ project, 61
Compile, 22
Control Panel, Windows, 11, 13-14, 106,
   161, 214

D

Datasheet, 232
   ATtiny85, 208, 220, 235
   DHT22, 94
   Maxbotix, 99
   PING))), 182

Servo, 168
SPI, 184
Winbond, 25Q80BV, 186
Debugging, 157
Device Manager, Windows, 13-14
DHT22 sensor, 92
Digispark, 205, 208
Digistump, 205, 208
Digital pin, location, 5
Display, 232
    Buffering, 86
    I2C, 85
    LCD, 88
    LCD test code, 90
Driver (see USBtinyISP driver)
    Signed, 12
    Troubleshooting, 15
    Unsigned, 223
    Windows 8, 12
DS1307 RTC, 111, 115

E

EagleCAD, 224
eBay, 231
EEPROM, 5, 197
    Library, 39, 197
    Write cycles, 197
Error
    4, -5, 219, 222
    PCREL, 220

F

F macro, 135
FastLED library, 40, 56
Field of view, 145
Flash memory chip, Winbond, 185
Floating-point numbers, 51
freeRam function, 51
Frequency, 34
Fried, Limor, 2, 207
FTDI Friend, 103, 112
F_CPU value, 96

G

Gammon, Nick, 108, 209
Gemma, 8, 122, 202, 208
GemmaBoot, 204
Getting help, 15
GitHub, 39, 42
Goggles project, 68
Google, 110, 209

Google Glass, 78

H

Handshaking (serial), 107
Headers, 18
Hex file, 204
Hookup wires, 18

I

I2C, 84-85
    Addressing, 84
    Backpack, 88, 232
    Display types, 85
    Master, 84
    Slave, 84
    Wire library, 85
Impedance matching, 35
Instructables, 205
int16_t, 52
int32_t, 52
int8_t, 52
Internet of Things, 1, 3, 86, 164

J

JST connector, 29, 123

K

Kaleidoscope goggles, 68

L

Ladyada (see Fried, Limor)
Latching, 143
LED, 3
    NeoPixel, 55
        (see also NeoPixel)
    Tricolor, 55
    Types, 56
LED Backpack library, 40
LED Color Organ, 61
Level shifting, 194
Libraries, 37-50
    Dependencies, 218
    Errors, 218
    Illegal characters, 218
    Installing, 41
    Limitations, 50
    Problems, 217
    Third-party, 39
    Use, 49

libusb-win32, 14
Limor Fried (see Fried, Limor)
Linker, 11, 219
Linux, 12, 14-15, 223
LiPo battery (see Batteries, LiPo)
Low pass filter, 194

M

Mac, 10
    Download issues, 10
    Library install, 47
    No connection erro, 216
    Security, 10
    Troubleshooting, 215
Maxbotix sensor, 96
Memory
    Flash, 4
    Heap, 56
    Optimization, 50-53
    Program, 134
    RAM, 4
Memory map, 4, 200
Meters, analog, 113
Monk, Simon, 207
Mounting, 233
Music
    Mathematics, 35
    Scale, 34

N

Narcoleptic library (see TinyNarcoleptic
    library)
NeoPixel, 3
    Connections, 64
    Library, 40, 73
    Limit, 224
    Memory, 51
    Power, 64
    Rings, 68

O

Ohm's law, 113
Optimization
    Program, 51
    RAM, 52
    Variables, 51
Oscilloscope, 19

P

Parallel port, not available, 216

Passive Infrared Sensor (PIR), 136
PCREL error, 220
Pebble, 78
Photocell, CdS, 30, 169, 171
Ping))) sensor, 181
pinMode function, 223
Potentiometer, 79, 81
Power connections, 5, 28
Power savings, 127
power.h, 39, 96, 124
Preprocessor symbols, 220
Processing language, 188, 192
    Learning, 236
PROGMEM, 134
Pull-up resistor, 27, 149
Pulse Width Modulation, 108
    Diagram, 109
    High speed, 185
    Pin location, 5
    Servo, 79
    Smoothing, 193
    Trinket pin 4, 109, 223
pulseIn function, 102
PuTTY, 106, 162
PWM (see Pulse Width Modulation)

R

R-2R ladder, 149
RAM, 4, 134 (see Memory, RAM)
    Optimization, 52
Raspberry Pi, 205
Reading, suggested, 11
Reset button, 5-6, 13-14, 21-22, 27, 122,
    128, 131, 133, 195, 202, 213-214,
    219-220, 222
Reset pin, 27
Resistor, 18, 232
    Kit, 19, 232
    Pull-up, 27, 84, 92
RGBLCDShield library, 40
RS-232, 102
RST pin, 5-6, 27, 202, 213

S

Schematics, 28
Screws, 233
SendOnlySoftwareSerial, 39, 108,
    112-113, 209
Sensing, 3
Sensors, 91, 232
Serial, 102-108

No monitor, 214
Not defined, 222
Object, 107
Servo, 78-83
    Analog feedback, 83
    Horns, 79
Servo library, with IR library, 219
Servo8Bit library, 41
Shield, display, 149
Show and Tell, Adafruit, 209
Sketch, 21
sleep.h, 39, 124
SoftServo library, 39, 81, 181
SoftwareSerial library, 39, 104, 107, 108
Solder, 18
    Learning, 12, 235
Soldering, 12, 18
Soldering iron, 18
SPI, 183
Splice, 71
stk500_getsync(), 219
Supply voltage, 7
Support, manufacturer, 224
Switches, 148

T

Teensy, 199
Temperature and Humidity project, 86
Terminal program, 106
Theremin project, 30
Timers, 38
TinyDHT library, 40, 88
TinyFlash library, 40, 188
TinyLiquidCrystal library, 40, 88, 90
TinyNarcoleptic library, 40
TinyNewPing library, 41
TinyRTClib library, 40
tinySPI library, 41
TinyWireM library, 39, 85-86, 88, 99,
    113, 124, 139, 218, 221
TinyWireS library, 39, 85
Torque, 79
Toy Animal project, 165
Trinket
    16 MHz mode, 96
    Adafruit support forum, 224
    Bootloader restore, 202
    Buy, 231
    Compared to Uno, 2
    Connections, 5
    Schematics, 28
Trinket Alarm project, 146

Trinket Audio Player project, 185
Trinket Jewelry project, 121
Trinket Occupancy Display project, 135
Trinket Pro, 231
Trinket Rover project, 173
TrinketHidCombo library, 40
TrinketKeyboard library, 40
Troubleshooting, 211
TTL level, 102
Two-Wire Interface (see I2C)

U

uint16_t, 52
uint32_t, 52
uint8_t, 52
Ultrasonic sensors, 96
Universal Serial Interface, 37, 84-85,
    183
USB, 12
    Bad cable, 211
    Hub, 13, 19
    USB 2 port, 19
    USB 3 issues, 13, 213
    USB 3 Issues, 19
    VID/PID, 201, 204
USBtiny device, 14
USBtinyISP driver, 11-15
USBtinyISP programmer, 201, 203

V

V-USB, 4, 201, 213
VCC, 7
VirtualWire library, 41
VMware, 214
    Mac, 216
Voltage divider, 32, 190

W

WAV files, 188
Wearables, 3, 78, 121
    Cosplay, 68
Windows, 10
    Control Panel, 11, 13, 106, 161, 214
    Library install, 44
Windows 8, 12
Winscot, Rick, 173
Wire library, 85 (see I2C)

Z

Zhao, Frank, 208

About the Author

Engineer, Maker, and innovator Mike Barela is currently a senior foreign service officer with the US Department of State. A graduate of Whitman College in mathematics/physics and the California Institute of Technology in electrical engineering, he has worked at Hewlett-Packard, the Caltech/NASA Jet Propulsion Laboratory, and Boeing. Mike has traveled the world, living in a number of countries providing security to American embassies. An avid electronics enthusiast, he has worked with personal computers since the introduction of the PC. He rekindled his electronics and micro-controller interests of late, authoring a number of popular articles on using Arduino-compatible systems. This includes collaboration with Adafruit Industries on a number of tutorials (*http://learn.adafruit.com*).

The cover and body font is Benton Sans, the heading font is Serifa, and the code font is Sans Mono Code.

CPSIA information can be obtained at www.ICGtesting.com
Printed in the USA
LVOW02s1457270515

440107LV00008B/48/P

9 781457 185946